MW00774036

"While the new atheists claim that science and Christianity are at war with one another, *Revealing God in Science* rebuts that popular and unfortunate narrative. This book contains the stories of numerous well-trained scientists who are also devoted followers of Jesus Christ. They come from different backgrounds, different continents, with different specialties in the various STEMM fields, but their message is consistent: science and the Christian faith are allies! This is an instructive, encouraging, and enjoyable book to read. Give it to your skeptical friends and family members."

—Kenneth Samples
Senior Research Scholar
Reasons to Believe

"Rumors say that science and faith are at odds. If there *is* a conflict, the dividing line runs right through the hearts of Christians who like science or through scientists who wonder about Christianity. Is there any hope of peace? YES!

In this book, a dozen scientists share their journey of how God was revealed through science. I was moved and encouraged by their stories, and I think you will be too."

—Darren L. Williams, PhD
Professor of Physical Chemistry
Ratio Christi Chapter Director

"*Revealing God in Science* does a masterful job of challenging the entrenched 'war metaphor' that unfortunately has dominated much of the discourse about the relationship of Christianity and science. In chapter after chapter, scientists representing a broad spectrum of disciplines demonstrate why the Christian faith is, in fact, a reasonable faith. God really does speak through two books."

—Kenneth Keathley
Senior Professor of Theology
Southeastern Baptist Theological Seminary

"The narrative that science and Christianity are at war is one of those myths that never seems to die. *Revealing God in Science* features a powerful collection of stories from a diverse group of Christians in STEMM. These scientists provide a global perspective that demonstrates the harmony between faith in God and the pursuit of scientific knowledge. This book is an excellent resource, and I highly recommend it for students and professionals in STEMM."

—Joseph R. Miller, DMin, PhD
Assistant Professor of Christian Worldview
Grand Canyon University

"Brilliant scientists are people—meaning that no matter how smart they are, they find themselves ensconced in the very same world that everyone else inhabits. In this endearing book, you'll get to meet honest individuals who know they're human beings, not just scientists. Each one shares a personal account of how they came to terms with the God of science. If you're interested to hear the beating hearts of those who understand complex math and other principles of the cosmos, then sit down with this volume and dare to enter into holy joy."

—Sarah Sumner
President
Right On Mission

"I think you'll be grateful for this collection of testimonies and reflections and impressed with the caliber of people associated with Reasons to Believe. One feature I especially appreciate is the way this diverse group—men and women of differing nationalities and ethnicities, many of whom are accomplished practitioners in STEMM fields—show themselves capable of serious and engaging thought on deep issues regarding the meaning of life."

—C. John "Jack" Collins
Professor of Old Testament
Covenant Theological Seminary

Dedication

To say that this book owes its existence to Dave Rogstad would be no exaggeration. His influence on Hugh Ross, starting in their days together at Caltech, played a crucial role in Ross's life. Dave encouraged Hugh to remain in the US, become involved in Christian ministry, and to establish Reasons to Believe. Its publishing arm brings you this collection of stories from its growing Scholar Community.

REVEALING GOD
IN SCIENCE

12 STEMM Scholars
Share Their Journey

GEORGE R. HARAKSIN II AND KRISTA BONTRAGER
GENERAL EDITORS

© 2025 by Reasons to Believe

All rights reserved. No part of this publication may be reproduced in any form without written permission from Reasons to Believe, 818 S. Oak Park Rd., Covina, CA 91724.

Cover design: Danielle Camorlinga
Interior layout: Christine Talley

Unless otherwise identified, all Scripture quotations taken from the Holy Bible, New International Version® NIV®. Copyright © 1973, 1978, 1984, 2011 by Biblica, Inc.™ Used with permission. All rights reserved worldwide.

Names: Haraksin, George R., II, editor. | Bontrager, Krista, editor.
Title: Revealing God in science : 12 STEMM scholars share their journey / edited by George R. Haraksin II and Krista Bontrager.
Description: Includes bibliographical references and index. | Covina, CA: RTB Press, 2025.
Identifiers: ISBN: 978-1-956112-07-8
Subjects: LCSH Religion and science--Christianity. | BISAC RELIGION / Religion & Science | RELIGION / Creation | SCIENCE / General
Classification: LCC BL240.3 .R48 2025 | DDC 201.65--dc23

Printed in the United States of America

First edition

1 2 3 4 5 6 7 8 9 10 / 29 28 27 26 25

For more information about Reasons to Believe, contact (855) REASONS / (855) 732-7667 or visit reasons.org.

Contents

Figures

Acknowledgments

In the days prior to the Internet, I (George) would drive up to an hour to Reasons to Believe (RTB) to sift through three-ring binders filled with Hugh Ross's white papers on the integration of sound science alongside a testable faith. At the time I was a pastor and college professor of philosophy and ethics. Little did I know that I would later be an RTB scholar and staff member myself. Krista Bontrager (then RTB Scholar Community director) and Fazale "Fuz" Rana (RTB president and senior research scholar) hired me as the Scholar Community manager. I later become the director. It's been a dream come true to work and minister alongside such admirable people. I'm truly thankful to Krista and Fuz for trusting me to help them grow and develop the RTB Scholar Community to over 200 members at the time of this book's writing.

Krista and I want to thank Fuz for listening to our excitement for this project and for being an advocate for having this book published by RTB Press. Thanks goes to several of our colleagues at RTB for their help in keeping the project going. Scholar Community program associate director Tracy Weldon jumped in at several points in the project to help with communication and coordinating details (because "big idea" persons can get paralyzed by the details). RTB senior research scholar Kenneth Samples reviewed the manuscript during a period where he was busy finishing his own book. Scholar Community program coordinator Jacob Rodriguez supported final touches on the manuscript. Heartfelt thanks go to RTB's editorial team for improving the book's flow and readability, especially editorial director Sandra Dimas, who was patient with me (George) and coached me through the process. Sandra, along with Joe Aguirre, Maureen Moser, Helena Heredia, Jocelyn King, and Karrie Cano devoted themselves to editing and fact-checking, preparing the index, and more.

Our deepest thanks goes to Emily Bontrager, Kathy Ross, and Diane, Steve, and Daniel Rogstad for their invaluable contributions to David Rogstad's chapter, which were instrumental in completing it after the author's passing.

A special thanks to the RTB Scholar Community members who contributed to this book project. They were more than willing to invest their precious time, amid full schedules, into an RTB book for the good of others. We sincerely hope and pray the stories revealed in this book will inspire and energize many others that take seriously the faith once for all delivered to the saints (Jude 3) to pursue STEMM vocations.

Rumors of War (and Peace)

George R. Haraksin II and Krista Bontrager

"Yes, there is a war between science and religion," declares Jerry Coyne, professor emeritus of ecology and evolution, University of Chicago.[1] The use of the *war metaphor* or *conflict thesis* to describe the relationship between religion (specifically Christianity) and science is entrenched in culture. The conflict, whether perceived or real, results in various types of barriers making the religious skeptical of going into STEMM (science, technology, engineering, mathematics, and medicine) fields or causing scientists to be dismissive of religious knowledge and skeptical of participating in a church community. The purpose of this book is to raise skepticism concerning the applicability of the war metaphor. The testimonies and research of the scientists featured in the following chapters may stoke rumors of peace in the relationship between science and religion. They resemble leaflets sent out to those who hadn't thought WWII was over.[2] To quote John Lennon, "War is over, if you want it."[3]

According to a 2015 Pew Research study, "73% of adults who seldom or never attend religious services say science and religion are often in conflict, while half of adults who attend religious services at least weekly say the same."[4] Whether or not the war metaphor is justified, in our contemporary cultural context there is little doubt that there's a *perception* of conflict between religion and science. But this has not always been the case.

Recent Development of the War Metaphor

In the nineteenth century, scientist John William Draper and historian Andrew Dickson White became wildly instrumental in manufacturing the so-called "war between science and religion." Draper, a chemistry professor at the University of New York, wrote *History of the Conflict Between Religion and Science* in 1847.[5] His book was unexpectedly influential and went through 50 editions. White picked up the mantle in 1896 by writing *A History of the Warfare of Science with Theology in Christendom*.[6] This fellow architect of the war metaphor was the founding president of Cornell University, an institution that set itself apart with its secular orientation.

One might have the impression that the two enterprises were already embattled when Draper and White came along. But historical scholarship reveals a different account. As historian Timothy Larson observes, "Draper and White were not simply describing an ongoing war between theology and science, but rather they were endeavoring to induce people into imagining that there was one."[7] Draper's background was in English Methodism and he was worried about the power wielded by the Catholic Church. He wrote from the perspective of a successful scientist with science having the final say and not the (Roman Catholic) church. White was president of Cornell University while writing his most important work. Cornell was the first nonsectarian university in the US and White's target were those who advocated for sectarian, theological education. Cornell was an open challenge to dogmatic theology and sectarian (theological) education. One way of invoking this image of conflict in people's minds was to concoct the story that the church opposed several new scientific breakthroughs and developments.

One such story was that the historic Christian church had maintained that the earth was flat. This claim is false. For millennia, people have known the earth was round. People observed an eclipse of the Moon and saw that Earth was round. The ancient Greeks knew it was round and made attempts to measure it. The people of the church were no different. The church fathers said Earth was round. Christians of the Middle Ages knew it was round. The renowned church doctors from Saint Augustine to Saint Thomas spoke of the earth's sphericity.

But the myth survives to this day. In the book *unChristian*, authors and researchers David Kinnaman and Gabe Lyons report on a contemporary church's sermon series, which includes a sermon of apology and repentance titled, "We're Sorry for Saying the Earth Is Flat."[8] While war metaphor and conflict thesis remain in popular culture and publications, current historical scholarship rebuts the notion that most notable Christians held to such a view.

David Hutchings, in his book *Of Popes and Unicorns: Science, Christianity, and How the Conflict Thesis Fooled the World*, writes, "In reality, hardly any Christian writers of note (most likely, no more than two) argued for a flat

Earth, and they were largely ignored."[9] Hutchings details how the conflict thesis or war metaphor has been thoroughly debunked in the literature and should be viewed mainly as myth. "Despite this," he observes, "they still circulate today, and many still believe that we must pick a side: God or science." Hutchings's account is "a history of science and religion, and of how, despite the common acceptance of the contrary, the latter has actually been of great benefit to the former. Rumors of a centuries-old war between God and science, it turns out, have been greatly exaggerated."[10] The life stories of the contemporary scientists in this work will hopefully help us to move beyond such flawed tales spun by Draper and White—among others—and realize that the relationship of religion and science, or, put another way, "faith and learning," can benefit one another and live in a creative collaboration that seeks truth for both theists and nontheists.

Quest for the Transcendent
Can the scientifically inclined still hold space for spirituality? Many nontheistic scientists struggle to shed what they're willing to call the sacred, inspiring experiences, or enchantment. Theologian and comparative religions scholar Rudolf Otto long ago termed it *numinous experience*—feelings of transcendence, something "wholly other," overwhelmingly powerful yet merciful.[11]

Physicist and self-described "spiritual materialist" Alan Lightman writes about his numinous experience in the book *The Transcendent Brain: Spirituality in the Age of Science*. Lightman's experience and "feeling of being a part of something larger than myself," occurred after a close encounter with charging juvenile ospreys coming at him like two F/A-18 Super Hornets. Ospreys have strong, powerful talons that can do much damage, yet Lightman held his ground and made sudden eye contact with one of the birds before it veered up in a dazzling, lightning-fast maneuver. Left shaking and in tears, yet with a feeling of connectedness to nature, Lightman admits that "to this day" he doesn't understand what happened in that brief moment, but he's willing to place the experience under the label of "spirituality."[12] As others have observed,[13] Lightman's spiritual materialism conflates science with a materialistic worldview, yet he also desires for science and spirituality to remain in dialogue and not in a duel. Given our increasingly polarized world, he writes, "the dialogue between science and spirituality has assumed greater and greater importance. Science and religion/spirituality are the two most powerful forces that have shaped human civilization. Neither is going away. Both are part of being human."[14]

Although we question Lightman's views in some areas, we can affirm with him the great importance and power of science and religion in our lives. The scientists that we encounter in this book—who hail from around the globe—reveal the part these two forces have played in their life journeys.

Meet the Scientists

To start our journey from "war" to peace, astrophysicist Jeff Zweerink engages both science and philosophy as he ponders whether science and Christianity are a match made in heaven. Next, Balajied Nongrum, the current lead consultant of research for Reasons to Believe Asia–Pacific (RTB APAC), wrestles with whether faith in God must make sense. Infectious disease physician Francisco Delgado narrates his understanding of spiritual experience, how experience is testable, and how to test science and faith. OB–GYN Christina A. Cirucci reflects on life and suffering—drawing from episodes in her own life and from time spent performing medical work in Bangladesh. South African astronomer David L. Block describes himself as a Jewish scientist who understands how fellow Jews wrestle with the claims of Jesus of Nazareth and the atrocities of the Holocaust. Biochemist Fazale "Fuz" Rana demonstrates how a scientist can see the Designer in the laboratory.

One does not usually see a connection between physics, faith, and archaeology, but John A. Bloom tells us of a quest worthy of an Indiana Jones sequel. Cynthia Cheung, a Hong Kong–born scientist with a career at NASA's Goddard Space Flight Center, discusses reasons why, historically, science was stillborn in China. Nicknamed "Rocket Girl" by her colleagues, Leslie Wickman reflects on her early religious journey and a diverse scientific career from working with NASA on the Hubble Space Telescope project, to R&D for the Bellagio Fountains in Las Vegas, to also playing professional tackle football and competitive volleyball. University of Oklahoma physicist Michael G. Strauss explains the compatibility between the big bang and the Bible. In the concluding chapters, we'll hear from two CalTech–educated scientists. The first is the late physicist David Rogstad, who worked on numerous projects—including the Galileo mission—with the Jet Propulsion Laboratory in Pasadena. Finally, astrophysicist Hugh Ross, founder of RTB, explains how several key people and events shaped his early life and career and how science affirms the message of the Bible.

Whether one identifies as a theist, scientist, spiritual-but-not-religious person, or something else, this book is an invitation to listen, learn, and live in that creative collaboration between science and faith and the truth and knowledge that result from their relationship. Contemporary science employs people from various viewpoints and worldviews: theists and nontheists; spiritual materialists and spiritual supernaturalists. We hope that as you read about the journey of each scientist, you'll sense a spirit of gentleness and respect as they demonstrate the harmony between science and faith.

Chapter 1

A Match Made in Heaven

Jeff Zweerink

I remember the conversation vividly. Our Sunday school class met in the foyer outside the sanctuary. Glass windows and doors overlooked the parking lot and cornfield beyond. As a new graduate student, I had the opportunity to meet many new people. This particular Sunday morning a fatherly fellow struck up a conversation with me. Eventually the topic turned to my work. I was telling him about my studies in astrophysics and how much I enjoyed science. After a few minutes of conversation, he looked at me with concern and care and commented, "At least you still have your faith."

Do you get the implications of that statement? According to my new friend, scientific studies ultimately and inevitably lead to the destruction of Christian faith. That's a tough stance, but I'm here to bring a far more hopeful message to Christians and skeptics who've bought the perception that science undermines faith and faith stifles scientific advance. I'm convinced that, instead of opposing one another, science and Christianity belong together. In fact, Christians should actively encourage and pursue scientific careers. Why do I make this claim? Because science and Christianity truly are a match made in heaven. Wrestling with doubts allows this match to flourish because the wrestling produces confidence—without self-righteousness.

Wrestling with Creation Passages

My first exposure to this match made in heaven came while I was an undergraduate. After my senior year of high school, I participated in a summer mission trip to Europe. There I learned how to do evangelism. During that summer my desire to walk with God grew tremendously. One obvious outworking of this growth was a desire to study the Scriptures regularly, which I did throughout my years in college. One semester, I was studying through Genesis. The footnotes of my Bible said that the days of Genesis 1 had to be 24-hour days. I believed in the authority and inerrancy of the Bible, so I figured if that's what the Bible says then that's what I believe. However, I wrestled with this idea quite a bit because my studies in physics seemed to indicate that the earth and the universe were much older than a few thousand years. Perhaps scientists had misunderstood time—maybe time was exponential instead of linear.

Then, in my sophomore year, a prominent Christian and scientist named Hugh Ross came to speak at Iowa State University. In his talk he claimed that the universe was indeed billions of years old and that the latest scientific measurements agreed with how the Bible described the universe. I didn't know how to reconcile the vast ages with my knowledge of the Bible. So, I asked Hugh that question when I met with him the next day. He explained that *yom*, the Hebrew translated *day*, had several different meanings, including "a long, finite period of time." And while many Christians interpret the days in Genesis 1 to say that the earth is 6,000–10,000 years old, other Christians say the correct interpretation is that the earth is much older.

I went to the university library for some fact-checking. Specifically, I looked in all the lexicons to see how they handled *yom*. Lo and behold, those lexicons validated what Hugh had said about different meanings for *yom*. I also learned that many respected Christians who hold the Bible in high regard differ in how they interpret the days of Genesis 1. What I learned during this conversation and subsequent research was that I didn't have a good understanding of how Christians properly interpret the Bible. Recognizing that the age of the earth wasn't a conflict between science and Scripture, I set about checking Ross's claim that big bang cosmology matched how the Bible describes the universe. And I found that he was right.

Discovering Harmony Between Science and the Bible

This approach to the Bible made a lot of sense to me. As I had read through the Scriptures, I noticed how it says in numerous places that God reveals himself in creation (e.g., Psalm 19 and Romans 1:18–21). In fact, according to Romans 1, as we look at creation, God's revelation makes it clear that he exists. So, when the Bible says, "In the beginning God created the heavens and earth," I would expect the study of creation to show that the universe had a beginning. This feature is one of the claims of big bang cosmology.

The Bible also describes creation's reliability and dependability. The prophet Jeremiah makes the statement that God will be trustworthy and reliable in keeping his promises. Specifically, the passage says that God's covenant with day and night and his establishment of the fixed patterns of heaven and earth reflect God's trustworthiness (Jeremiah 33:25–26). Scientifically speaking, the laws of physics govern how heaven and earth behave and how day and night happen. So, we might say that the constant laws we observe governing the universe give us great confidence to expect that God will keep his promises. It may surprise you to know that the philosophical motivation for Einstein to develop his theories of relativity was the idea that the laws of physics ought to be constant throughout the universe.[1] I can tell you this: every experimental test that scientists have thrown at the theory of relativity has passed with flying colors. We live in a universe governed by constant laws of physics.

Another biblical description of the universe states that God is stretching out the heavens (Job 9:8, Isaiah 42:5, Zechariah 12:1, and others). I'm not sure that the biblical authors had the expansion of the universe in mind, but I find this language to be provocative considering big bang cosmology!

If we were to ask what scientific model has constant laws governing a universe that begins to exist and expands, the answer is big bang cosmology. This model, endorsed by the majority of scientists, corroborates the biblical description of the universe.

I could also talk about the design that scientists see as we investigate the universe—how the universe has a purpose to support humanity. We could talk about the exquisite design of the earth and its ability to support life. We could talk about the evidence pointing to an original pair of humans, or the intricate design of the genetic code, or what it takes to make a universe with abundant carbon and water, or how humans are the only creatures that seek to worship God. All these features of the natural realm point to a remarkable truth: what we see in creation and what we see in the Bible agree. They go together, they belong together. Science and Christianity really are a match made in heaven.

Resolving Doubts
I don't want to imply that I haven't had doubts as I've studied creation and read the Bible. In fact, I recall a time after I finished graduate school when I had studied a lot of science apologetics. I had gone through the training at Reasons to Believe to become a volunteer apologist. I answered numerous questions from other apologists. If you asked me directly, I would say, "Yes, of course the Bible and science agree." Yet as I was reading about new scientific discoveries, I began to feel anxious. And it wasn't a one-time thing. Reading about the

validity of evolution, life on Mars, and other challenging discoveries increased my anxiety.

Spending time in reflection and prayer revealed a disturbing thought. With each new discovery, I worried that perhaps this was *the one* that would show Christianity to be false. However, my anxiety provided me with a great opportunity—the chance to think rightly about science and the Bible. If God is the creator of the universe (I believe he is) and if he inspired the Scriptures (which I believe he did), then the scientific discovery that disproves the Bible will never happen. That realization relieved virtually all my anxiety.

But it didn't remove all my doubts. I encountered my first big doubt a year or so later.

The Multiverse Challenge
In the early years of the millennium, scientists were just starting to talk about the multiverse—the idea that our universe is one among a vast multitude of universes. In 2001, I encountered the challenge of the multiverse at a conference dinner, during which I engaged in one of the more interesting discussions about science and Christianity I've ever had. About 10 scientists, myself included, were gathered around the table. The topic of how science might investigate the existence of God came up, and we discussed it for about 45 minutes. We talked about big bang cosmology and the fine-tuning of the universe. I commented that these features pointed toward a Beginner and Designer of the universe. Then one of the prominent scientists in the group looked right at me and asked, "That's all fine, but which universe are you talking about?"

I had been told that the multiverse had no scientific legitimacy. But here was a prominent scientist using it to rebut my claim that big bang cosmology pointed to God. Immediately, I saw the implications of the multiverse, and didn't have a good response. Inflation, a key component of big bang cosmology, can make many universes. Although our universe had a beginning, maybe there's nothing significant about that beginning—kind of like how Sunday is the beginning of the week but is not anything fundamentally important. Perhaps inflation had been making universes forever. And while our universe may *look* designed, the tremendous size and variability in the multiverse meant that something like our universe existed somewhere. Our existence places exacting conditions on the universe, but the fine-tuning was unexceptional because it happened "naturally."

Following that dinner conversation, I wrestled with the evidence for a multiverse and its implications. For about a year, I didn't have a suitable response. I should note that my wrestling was not about the truth of Christianity, but with using big bang cosmology to argue for the cosmological and teleological argument. Still, the doubt and frustration I experienced was real. Finally, after my year of wrestling, I discovered something profound. First, scientists have good reasons for thinking that a multiverse exists. It wasn't just some idea they found attractive to avoid the existence of God. More importantly, continuing research into multiverse models revealed that even the multiverse had a beginning and our being here demonstrates evidence of design or purpose. In other words, while the multiverse initially appeared to undermine important arguments for God's existence, more research ultimately made the arguments stronger and more robust. I won't go into the details here, but would encourage you to check out *Who's Afraid of the Multiverse?*[2]

The Challenge of Extraterrestrial Life
Without a doubt, the other big issue that raised questions in my mind was the possible existence of life beyond the confines of Earth. But in wrestling with this question, I could apply the valuable lessons I learned while struggling with the multiverse. I knew how to make a strong argument that the chances of finding another habitable planet like Earth were beyond remote. But a nagging question rattled around in my mind—What if scientists actually find life out there? If Christianity demands that Earth is the only place that supports life, then the discovery of extraterrestrial life would falsify the faith.

This question surfaced as I read papers showing that the thousands of discovered exoplanets meant that hundreds of billions of planets the size of Earth exist in just the Milky Way Galaxy alone. And other papers claimed that 20% of stars have potentially habitable planets.[3] Moreover, the universe seems to readily produce carbon, hydrogen, oxygen, and nitrogen—the stuff that life is made of. Might then it be possible that life exists somewhere out there? As I thought about this question, I applied my multiverse lessons. First, I investigated what we know about exoplanets. It turns out that our knowledge base is confined to a rather limited set of information. All we really know about these exoplanets is their orbits, masses, sizes, and, on rare occasions, temperatures. At this point, we have no idea whether any of these "potentially" habitable planets even have liquid water on them—and we won't know for at least the next couple of decades.

Then I investigated what the Bible teaches on this issue. I found a lot of passages explaining what God did here on Earth and how he revealed himself to us. But very few biblical constraints exist regarding life beyond Earth. In fact,

I discovered that Christians through the centuries have thought extensively about this question and have found many ways that extraterrestrial life would fit comfortably within Christian doctrine. Furthermore, many scientists who are devout Christians come down on different sides of the question. Some (like Johannes Kepler) argued that God created life on numerous planets. Others (like Galileo Galilei) argued that God created life here on Earth only.[4] This helpful history of thought means that I can use this topic to engage people in gospel discussions without having to first convince them that we are the only life in the universe.

At the end of the day, my wrestling with these topics has strengthened my confidence in the truth of Christianity. However, there's one more important lesson I've learned. Let me tell you a bit more of my story.

Doubting My Father's Love

Two of my earliest memories relate to my dad. When I was three years old, I was sitting at the top of the stairs in our duplex watching my dad perform a set of science demonstrations for a group of my older brother's friends. He would dip a tennis ball in liquid nitrogen and then shatter it on the wall. He would also mix chemicals together, and they would make funny noises or turn bizarre colors or fascinating shapes. Some of the demos seemed like magic. At that young age, my admiration for my father seemed boundless. I have seen these demos a hundred times and would watch them again if given the opportunity. For as long as I can remember, I've been fascinated by the way things work and wanted to understand.

A year or so later, I remember seeing my parents baptized in the 102 River outside St. Joseph, Missouri, where we lived. My parents' Christian faith permeated our home. We regularly attended church, served others, learned to pray, and read the Bible. One of the most dramatic influences, though, was how my dad and mom desired to love us the way Christ loved them. As they taught us Christianity, I came to know Christ as my personal savior during an AWANA program in the fifth grade. I don't remember the specific message, but during one lesson time, I recognized that God loved me and Christ's death on the cross would reconcile me to him if I would accept his gift. So, I did. My shyness played a part in the story because the speaker encouraged us to tell our leader so that they could talk more with us. I didn't do that, but I did tell my parents that night. My dad assures me that I understood the gospel and trusted Christ that night.

The pivotal lesson came in the context of an annual fishing tradition that my dad, brothers, and I started during college. We would head down to a lake in southern Missouri or float a river while enjoying the thrill of hooking the big one! Over the years, the fishing trip grew, as did our family bond.

Eventually, we set our sights on the Gulf of Mexico. I couldn't wait! However, this particular year, my dad planned the trip at a time when I couldn't go. My wife was pregnant with our second son. She was due to deliver less than a month after the trip and was experiencing preterm labor difficulties. As I assessed the situation, here was the person I respected and admired most planning an important tradition and leaving me out. Why didn't he reschedule? I was fuming! How could he truly love me and do this?

The trip put me in a bind. Although I was angry, I knew that after the trip I needed to talk to my dad and see how it went. It crossed my mind that it might be easier to dramatically limit or sever my relationship with my dad rather than deal with the problem of leaving me out of the great family tradition. Fortunately, I had a month. As I started praying about how to respond, God prompted me to dwell on the truth. I was angry with my dad, but I began to remember how my dad had regularly taken time out of his busy schedule to coach my soccer team. He taught me to build LEGO, let me win at chess, and traveled to all my high school football games. One of those games stood out. My dad taught chemistry at the college in my hometown and had earned a teacher-of-the-year award. Through my mom I discovered he was planning to go to my run-of-the-mill weekly football game *instead* of attending the award ceremony. I let my dad know that he should go to his award ceremony. But I received the message loud and clear—my dad loved me!

As I recalled these events, I had a choice. I could focus on the hurt and anger I had because of the missed fishing trip, or I could choose to believe my dad loved me based on clear evidence from my childhood. I can tell you this. Nothing about the circumstances changed. I missed the fishing trip and never really got an explanation about the timing until years later. But as I chose to believe the truth, my anger and resentment began to melt away. Furthermore, despite challenges like the fishing trip—no, actually *because* of them—I've chosen to believe the truth, and my relationship with my dad continues to grow stronger each year!

This encounter taught me a valuable lesson. My dad loved me regardless of what I thought, but my *experience* of his love required me to make choices. When challenges arose, I could choose to let my feelings drive my response. If so, the pain, doubt, and uncertainty would define my relationship with my father. Alternatively, I could focus on the truth and let the facts, evidence and stability define the relationship. Fortunately, we can all apply that foundational lesson when we encounter doubts about God.

Doubts Are Not Necessarily Bad

We've all had doubts. I have, you have, and you know people who have doubts and questions. So, how do we help someone wrestling with doubt? Jesus gives us a great example to follow in Matthew 11. When John the Baptist was in prison he sent his disciples to Jesus with this question: "Are you the one who is to come, or should we expect someone else?" (Matthew 11:3). If you think about this query for a moment, you'll realize that there's not a more ludicrous question. While still in the womb, John recognized Jesus. John's sole mission in life was to prepare the way for Jesus. He baptized Jesus (after briefly protesting about his own unworthiness) and then saw the heavens open, the Spirit of God descend on Jesus, and a voice from heaven proclaim, "This is my beloved Son, with whom I am well pleased" (Matthew 3:17). If anyone should have known who Jesus was, it was John the Baptist!

How did Jesus respond? First, he welcomed the honest question. Jesus didn't say, "Really, John?" or "Come on, John, you just gotta have faith." Jesus was not offended. Second, Jesus did not preach a sermon for John to believe or give a data dump of everything he knew that might relate to John's question. Jesus responded simply, "Go back and report to John what you hear and see" (Matthew 11:4). Third, Jesus gave John a concise answer that addressed John's deep need to know and believe the truth—and then he let John decide how to respond. I find this third step the hardest challenge since it requires me to know the person well, to know the evidence well, and to relinquish control to God.

This process will play out over and over for those who study science. Many people claim that the challenges raised by studying nature undermine belief in Christianity. Some assert that science and faith are in conflict and science will emerge triumphant. But the evidence paints a different picture. Science and Christianity truly are a match made in heaven. And though we will wrestle with doubts, we'll resolve them, and our confidence in the truth of Christianity will grow. Then we'll be even more equipped to share God's love with conviction, grace, and humility—just as God commanded us!

Chapter 2

Finding Answers in Christ

Balajied Nongrum

I grew up in India, a nation known for its diverse cultures, foods, languages, and faiths. It's said that within this great nation you'll discover several different nations on account of the various ethnicities and cultural expressions. And though India as a whole is highly religious, I learned early on that faith in God (or gods) is far more subjective than objective. However, I am convinced that in our quest for the truth, the subjective approach to faith ought to be anchored in the objective truth claims of a worldview supported by concrete evidence.

In my effort to search for meaningful confirmation of whether God exists and is indeed the Creator, I have engaged with various questions that have been used to challenge the truth. Such questions range from the origin of the universe to the problem of pain, evil, suffering, and injustice. Observing science has helped me to reason that God's position as the Creator cannot be dismissed. Science reveals the complexity of life. At the same time, science cannot entirely disprove God's existence. Even problems that arise due to pain and suffering can be understood through the validity of answers that come from having faith in God.

Therefore, through careful reasoning and handling of evidence from science and Scripture, I make the case that my (subjective) faith in God is backed by objective and verifiable reasons, and that belief in God makes the best sense of all life's big questions.

The Road to Discovery

I was twelve years old when I was admitted into a boarding school. Although I was raised in a Christian home, my understanding of my belief in God was very shallow. Through my interaction with friends from other religious backgrounds, I learned about various views on life, cultures, and faiths. One view that seemed to reign supreme among my peers at the time was that all religions are ultimately the same, but they look different superficially. They would argue that we are heading toward the same destination via different paths (or faiths). That view influenced me, and for many years (prior to my conversion) it was my default worldview.

The world of science had long fascinated me. My father, a physician, would often call me and my siblings together to explain to us the mechanism of action or the functions of human organs. His insights on human physiology intrigued me. He also encouraged us to take up science in our academic pursuits and to read widely and consistently. His encouragement served as a breeding ground for me to think rationally and objectively, which led me to pursue science in college with a focus in physics, chemistry, and biology.

I did well in my studies, but in the midst of academic pursuits, I went through an emotional and intellectual crisis. A deep sense of emptiness plagued me. My life lacked meaning and purpose. "What is the point of it all?," I wondered. To fill the void, I even pursued hedonism. But the more I plunged myself into it, the emptier I felt.

It was during this time of crisis that I met a man named Prosperly Bell Lyngdoh, who discussed the big questions of life with me. Eventually, he introduced me to the person of Christ. I was intrigued. As I further examined Christ and his truth claims, I discovered four unique features of the Christian faith that resonated deeply with me.

First, Christ gives me hope (John 3:16) for the future. This hope is ultimately anchored in his resurrection from the dead (1 Corinthians 15). Second, he grants me rest (Matthew 11:28) and relief from the burdens (and the emptiness) that were weighing heavy upon me. These burdens had often robbed me of the joy of living. Third, he offers me forgiveness (1 John 1:9) from all the guilt that I was carrying from the past, thus enabling me to start afresh. Finally, and most importantly, salvation "is the gift of God—not by works, so that no one can boast" (Ephesians 2:8b–9). Salvation is freely available to me and to anyone who is willing to receive it, and, above all, it is assured. These reasons were sufficient for me to commit my life to Christ.

The Call to an Adventurous Faith
Following my discovery of who Christ is and what he means to me, I was privileged to be admitted into university for my higher studies in veterinary science and animal husbandry. Again, I found myself discussing religion with

my friends. But while my boarding school peers' questions and conversations had centered on different faiths and their meaning and purpose, in a university that focused on science, the questions were more pointed. I vividly recall a question someone asked: "Hasn't science rendered faith in God redundant?" On another occasion, a peer of mine sincerely confessed, "I am just waiting for science to invent a cell and once science achieves this task, I will make science my religion!" Others asked me questions pertaining to the Christian faith such as, "What about the injustices in this world? What does Christianity have to say to the problem of pain and suffering?" and "In what way does the resurrection of Christ matter to our lives?"

Back then, with almost no in-depth knowledge about the Bible (or Christian theology) these questions and others baffled me. In my earnest desire to answer these questions, I began to study Christian apologetics. I devoured C. S. Lewis's book *Mere Christianity*. Lewis used easy-to-understand language to demonstrate brilliantly that the Christian faith is not only reasonable but also plausible. Simultaneously, I came across *The Complete Works of Francis A. Schaeffer: A Christian Worldview*. I read all five volumes. Schaeffer masterfully explained most of the essential Christian beliefs and their scriptural basis. These two thinkers, along with the context of a university campus, motivated me to start a small study group for students for the purpose of knowing what it is we believe and why.

Following my graduation from the university, I sought to discuss issues pertaining to the Christian faith and its relevance with people (both young and old) from different backgrounds. The insights I gained through these engagements (which were predominantly pluralistic in nature) gave birth to a desire to go deeper into the study of world religions. Hence, I studied philosophy and religion at Madurai Kamaraj University in India. I discovered that despite the popular belief that all religions are essentially the same, the truth is that all religions are fundamentally different and unique. This notion implied that every truth claim is necessarily exclusive.

Faith Through the Lens of Science

I had grown up with the firm conviction that science had nothing to do with faith (or religion), for they are worlds apart and always in conflict. Andrew White's *A History of the Warfare of Science with Theology in Christendom*,[1] published in 1896, contributed toward the conflict model. White contends that science has done away with God because science seems to have an explanation for everything in this world. Also, naturalism is the only possible framework available to interpret reality. Any attempt to reintroduce the biblical framework for interpreting reality is likely to be seen as a hindrance to the progress of science. Thus, the general perception is that all who pursue the scientific enterprise are naturalists.

This view, my graduate studies informed me, is entirely false. With the questions I had encountered in university lingering in my mind, I decided to pursue a master's degree program on science and religion via Biola University (Southern California). In doing so, I discovered the works of other academicians and philosophers of science who critiqued the mythologization of the relationship between science and the Bible. Nancy Pearcey and Charles Thaxton assert in *The Soul of Science* that the relationship between science and the Bible has been one of alliance rather than opposition. They corroborate this harmony by tracing it from the rise of modern science in the 1500s up to the 1800s.[2]

I learned from my studies that science is founded on certain philosophical assumptions: the principle of causality and the principle of uniformity. These assumptions are metaphysical (meaning "beyond the physical") in nature and have priority over all scientific investigations. Therefore, both these principles or philosophical assumptions ought not to be violated as we explore different scientific methods to answer questions about the origin of the universe or the origin of life.[3]

So, the question for me was, "How does the principle of causality help us connect science and faith?" The principle of causality states that every event or effect has an adequate cause. The Bible begins with the statement, "In the beginning God created the heavens and the earth" (Genesis 1:1). Is it reasonable to believe that God created the universe and all that is in it?

In the past, many people rejected this biblical statement as an argument for creation because they thought the universe is eternal. However, this idea was put to rest in light of scientific discoveries like the big bang and the expansion of the universe. Georges Lemaître formulated the modern big bang theory in 1927, and today, thanks to images of leftover radiation taken by the COBE satellite, it is an established scientific fact that the universe had a beginning. And if it had a beginning then there must also be a cause. This discovery puts Genesis 1:1 in a different light.

However, naturalists immediately raised an objection. They argued that the very claim of creation is false on the ground that its source is the Bible, and therefore it lacks any scientific basis. They would further assert that science can discover natural mechanisms for how things work while creation appeals to a God who cannot be detected.

Framework for Investigating Nature

The Bible is not a science textbook. Nevertheless, I realized that it touches on matters that are of great interest to science, including the origin of the universe and the origin and complexity of life. The Bible teaches that God is the author of Scripture (one source of revelation) as well as nature (a second source); thus implying that there are truths that can be obtained from both revelations. Therefore, this affirmation leads to the conclusion that there should not be any conflict or contradiction between the two sources of revelation.

Philosopher J. P. Moreland responds to the "creation is false" objection in his book *Scaling the Secular City* by pointing out that the objection commits an informal logical fallacy called the "genetic fallacy." The thrust of the argument is focused on the source of the claim, the Bible, rather than the claim itself—that is, the creation event. Moreland goes on to argue that the claim should be evaluated based on the evidence it offers for its validity and not just its source alone.[4] If the biblical account is indeed true, then both Scripture and nature are two of God's dynamic channels of self-disclosure.[5] Both these revelations should ultimately converge and reveal the same truth since "all truth is God's truth."[6]

Making Sense of the Origin of the Universe

Furthermore, what I learned about the word *creation* as used in the Bible is that it connotes the dynamic relationship that God (the Creator) has with his creation. He has created with a purpose. This idea is different from the scientific usage of the word *mechanism*, which, from a scientific point of view, deals only with how the universe runs and not with questions regarding its purpose. It affirms that the primary cause is outside the space-time domain and, therefore, can't be tested.

Hebrews 11:3 declares that "the universe was formed at God's command, so that what is seen was not made out of what was visible." This verse points out that the Creator, who is the primary cause for the cosmos, is "outside" the cosmos. Forms of multiverse theory seek to find causes of the universe from outside the cosmos, but such efforts are, by definition, not testable. If we look for causes inside the universe, we can find only secondary or efficient causes. In this sense, science can never prove nor disprove the existence of God, the primary cause.

Accounting for the Complexity of Life

The Bible provides a reasonable account for the complexity of life. Psalm

139:13–16 declares that we are fearfully and wonderfully made by God. When we consider the microcosm of life in a cell, we see how it testifies to the veracity of this claim. When we consider the complexity of information found in a cell and further examine its functions, we can't help but compare it to machines that are designed by intelligent minds. Similarly, when we look at the complexity of life in general, we can't help but conclude that this elegance in complexity would inevitably point us to an infinitely intelligent Mind.

Again, the naturalist may object and say that this biblical account of complexity isn't a mechanism, thereby rendering it invalid. However, energy, time, and chance alone can't produce the level of specified complexity that scientists observe. The additional inputs of intelligence and an ordering mechanism are needed to specify the information content. Thus, naturalism seems to fall short in explaining the complexity of life.

Why Evil and Suffering?
The problem of pain and suffering brought on by moral choices is a formidable challenge, but it doesn't necessarily lead to the negation of God's existence. On the contrary, I hold the view that God presents an answer to this conundrum by offering a vision of hope that seeks to make sense of evil and suffering. The Bible contends that God isn't merely an entity that exists; he embodies the attributes of being all-good and all-powerful. Scripture doesn't deny the reality of evil. Instead, it explains that God will address this problem and one day bring evil and suffering to an end (Revelation 21:3–4).

It would be presumptuous for finite human beings to claim to know God's purpose and design for why he would permit pain and suffering to exist. Yet, personal experiences and lessons from Holy Scripture can help us understand the possible reasons that God might have for allowing evil and suffering in his creation. This is the domain of theodicy, which is a branch of philosophical and theological conceptualization that seeks to justify the existence of a good, benevolent, and all-powerful deity despite the world being morally imperfect.[7] Many thinkers have addressed the problem of pain and suffering from this philosophical perspective, but I would like to offer a scientific take on trying to ascertain the purpose of pain in particular (with thoughts on evil and injustice to follow).

Pain: Foe or Friend?
Our modern world often views pain as an experience to be eradicated at all costs. At the slightest sign of a headache or any other pain, we gulp down

analgesics or painkillers. Medications that promise to remedy many of our ailments are readily available, a fact that highlights how science has played a key role not just in helping us deal with pain but also in attempting to eliminate it.[8] However, I think that ridding ourselves of all pain may not ultimately be good. There is good intention in the desire to alleviate pain because we have heartfelt empathy toward the sufferer. But I believe that pain and suffering may indeed play a vital role in our lives.

Consider patients with diseases such as leprosy, congenital painlessness, diabetic neuropathy, and other nerve disorders. Their inability to experience pain leads to greater harm to themselves than the disease itself.[9] From a scientific perspective, pain serves as a warning for the potential dangers that may lie ahead.

In the book *The Gift of Pain*, world-renowned hand surgeon Paul Brand and award-winning writer Phillip Yancey tell the story of a fourteen-year-old girl named Tanya, who was brought into a hospital with a swollen left ankle. On investigation, Brand was shocked to find that the "foot rotated freely, the sign of a fully dislocated ankle" and yet Tanya did not seem bothered at all and did not exhibit any pain.[10]

The rare genetic disease that Tanya had is informally referred to as congenital indifference to pain. According to experts, her only issue was that despite being completely healthy, she didn't feel any pain. Tanya "lacked any mental construct of pain" and had no "built-in warning system" to defend her from further injury.[11] This case led Brand to say that Tanya's condition:

> Dramatically reinforced what we had already learned from leprosy patients: pain is not the enemy, but the loyal scout announcing the enemy. And yet—here is the central paradox of my life—after spending a lifetime among people who destroy themselves for lack of pain, I still find it difficult to communicate an appreciation for pain to people who have no such defect. Pain truly is the gift nobody wants. I can think of nothing more precious for those who suffer from congenital painlessness, leprosy, diabetes, and other nerve disorders. But people who already own this gift rarely value it. Usually, they resent it.[12]

This fact made me reevaluate my own visits to the dentist, where the pain of having my decaying tooth repaired led to a better outcome. It kept me from having to suffer greater pain in the future. I can't help but agree with Brand's conviction that pain truly is one of God's greatest gifts to us, a gift that perhaps none of us want yet none of us can do without! (For more on this complex topic, please see chapter 4.)

Why So Much Injustice?

Next to the question of pain and suffering, perhaps the most vexing question we all grapple with is, "If God exists, then why doesn't he do something about the injustices we see everywhere?" But before we answer the question of injustice, we need to understand what justice means. C. S. Lewis wrote, "Justice means much more than the sort of thing that goes on in law courts. It is the old name for everything we should now call 'fairness'; it includes honesty, give and take, truthfulness, keeping promises, and all that side of life."[13] Even children from a very early age learn to speak this language of "fairness" whenever they're not treated equally. Humans seem to be wired with a strong desire for this world to be in good order. In other words, our desire for justice seems to be intrinsic to who we are. Yet, with the prevailing injustices that we see all around us, the longing for justice might be an exercise in futility.

I've often wrestled with the question of whether the world we live in is evil or not. It's true that not everything we see is evil; there are many manifestations of goodness as well. Good seems to coexist alongside evil. Could we suggest then that this is an evil world that's becoming good? Or, conversely, is it a good world that's going bad?

In answering such questions, we must consider ultimate questions like the origin, meaning, and purpose of life. Is this world designed by a Creator as the Bible declares, or is it the result of an accident, as naturalism affirms? If it's designed, then God is the reference point for all true justice. On the contrary, if it's merely an accident, then humanity becomes the ultimate reference point for all judgments. True justice in any society is one that is anchored in objective moral values that don't change either on the basis of time or culture.[14]

It's only after we base our judgments on such a foundation as an objective moral framework that we can meaningfully judge between right and wrong action, or between justice and injustice. Ultimately, an objective moral framework points to the existence of a moral lawgiver who is holy and righteous in character. Unfortunately, in our effort to make sense of (or to solve) the issues of injustice, we rarely consider how God will ultimately solve this problem. The psalmist says: "The Lord is known by his acts of justice . . . God will never forget the needy; the hope of the afflicted will never perish" (Psalm 9:16–18).

The Most Essential Truth Claim: the Resurrection

The reliability of the Christian worldview boils down to one single truth claim—the resurrection of Christ. According to the Bible, the truth of Christianity rests on the bodily resurrection of Christ. The apostle Paul wrote to the church in Corinth that, "If Christ has not been raised, our preaching is useless and so is your faith" (1 Corinthians 15:14). If Christ was indeed raised from the dead, then the resurrection is the greatest event in all of history. If Christians can dispel doubts and establish the veracity of accounts that support this truth claim, then we have compelling evidence for the truth of the Christian faith. The reliability of our faith in God revolves around this essential Christian truth claim that in turn is rooted in a historical context found in 1 Corinthians 15.

Two important characteristics about 1 Corinthians 15 can be pointed out when it comes to the evidence for the post-resurrection appearances of Jesus (1 Corinthians 15:3–5). First, this passage is one of the earliest reports available on the resurrection, even earlier than the Gospels. Second, these are eyewitness accounts. In fact, Paul recorded these accounts during AD 55—just 25 years after the crucifixion of Christ.[15] The short period of time between the occurrence of the event and its writing demonstrates the reliability of the eyewitness reports by ruling out the possibility of legends corrupting the report. As a basis for comparison, today we trust eyewitness accounts of events for which much longer periods of time have transpired in between.[16] Historical evidence is just that—evidence—and it favors the empty tomb and resurrection of Jesus.

Why the Resurrection Matters

We may not appreciate how hard it was even for the first disciples of Christ to believe in his bodily resurrection. Jesus's post-resurrection appearance was received with a degree of skepticism, so much so that Thomas, one of his closest disciples, said "Unless I see . . . I will not believe" (John 20:25). Seeing was believing for Thomas and probably for the others too. However, Jesus commended those who believed without such firsthand evidence.

Christians and skeptics today might ask a similar question: "What difference does it make whether we believe the resurrection is true or not?" Michael Green answers this question in his book *The Day Death Died* and offers three valuable insights. One, the resurrection of Christ tells us that God exists. This God—the one who brought this entire universe into existence—is revealed to mankind in the person of Jesus. God not only created us but loves us so much that he came to dwell among us, demonstrating to us that Christ is "the way and the truth and the life" (John 14:6). Two, the resurrection of Christ tells us that there is hope for our future. This life is not all there is, and death is not final. Believers in Christ will be raised just as he was raised. Three, the resurrection of Christ tells us that the evil and suffering we endure in life will end. We suffer for "a little while" (1 Peter 5:10), but we will experience glory for eternity.[17]

Green explains that Christ, through his resurrection, has once and for all shown that he is indeed the bridge between God and man. Christ's resurrection reveals to us that there is hope beyond the grave. The bodily resurrection of Christ provided the down payment for the world that is yet to come—a world without sin and death that we could only dream of but that we were utterly incapable of bringing about.

The book of Job describes how Job suffered greatly. He didn't find comfort in all the intellectual responses he received, but rather in his encounter with God. Job later testified: "My ears had heard of you but now my eyes have seen you" (Job 42:5). Similarly, Christ's resurrection enabled the disciples to come to the conviction that to see Christ was to see God.

Christ's bodily resurrection shows us that God is not only willing but also able to give us new life. The resurrection became the basis of the apostles' preaching as they invited the world to embrace the new humanity with all its promises in Christ. This event in history was the day that death died, and Christ invites all of us to believe and share in his eternal life.

A Testable Faith

Over the years, I've learned three valuable lessons. First, belief (or faith) in God is not purely subjective, and it has an objective reality to it. Second, all faiths (traditional or otherwise) are fundamentally different, though they may look the same superficially. Third, there's overwhelming evidence that makes belief in the God of the Bible reasonable. These evidences are objective and available for anyone to examine.

What fascinates me the most about the Christian faith is that it is testable. It invites you and me to think critically about what it teaches (Isaiah 1:18). I find this invitation refreshing because I can discover for myself that what I believe to be true in my heart is also true in my mind. In other words, belief in God does indeed make sense.

Chapter 3

Testing Science, Faith, and Spiritual Experience

Francisco Delgado

GOD of Abraham, GOD of Isaac, GOD of Jacob
not of the philosophers and of the learned.
Certitude. Certitude. Feeling. Joy. Peace.[1]

—Blaise Pascal, "Memorial"

I was born and grew up in Mexico, a historically Roman Catholic country whose culture was greatly shaped by the influence of the Catholic Church. My family observed some of the Catholic traditions, but religion did not take a central role in our family. We had a voluminous Bible at home that was seldom opened. People regarded the Bible as a book that was difficult to understand, and Catholic priests discouraged the reading of the Bible at home without a priest being present.

I called myself Catholic, but this just meant that I grew up in a Catholic country and in a Catholic family that followed Catholic traditions. Religion and religious ideas were discussed only at church on Sundays and at baptisms, weddings, and funerals. Even at these events the topics of discussion were rather repetitive. I don't recall hearing any sermons addressing topics of science from the pulpit.

I saw religion as belonging to the church and science to academia. The Mexican government and Catholic Church have had an uneasy relationship since the nineteenth century. The education curriculum was (and to my knowledge still is) controlled by the government. The educational system allowed for private Catholic schools, but both Catholic and nonreligious schools had to follow the official government curriculum, which embraced a naturalistic worldview. I attended a secular private school and, as I went from junior high school into high school, my interest in science grew. The incredible vastness of the universe, the predictability of the physical world, and the beauty of biology were mesmerizing. At school science topics were presented from a purely naturalistic perspective. No thought was given to the possibility of supernatural events. In fact, religion was rarely, if ever, discussed in school. Hence, my knowledge of world religions was superficial. Because of the sharp segregation of religion and science, I grew up with the belief that science and religion had no common ground.

A Guilty Conscience

I was an excellent student and earned good grades, but there was a dark side of my life unknown to almost everybody around me. As a teenager, I became involved in things that can only be described as evil. I had the sense that I was involved in wrongdoing, but the pull of the culture on my untethered human nature was very strong. Despite my desire to leave those things behind, I found myself unable to break free. I had been taught that confession at church, the pardon of a priest, and penitence would cleanse my guilt and change my heart, but despite going through the prescribed prayers multiple times, I found myself back in my old ways.

If God existed, I thought, then this God demanded perfection and would not look favorably upon the wicked nor hold back on his punishment. I deserved a lot of punishment for my wickedness, so I concluded that my wrongdoing was too evil for God to look favorably upon me. And if God would *never* look upon me favorably, then all I was left with was punishment. For this reason, I kept my distance from God.

My University Experience

As high school graduation neared, I started thinking about my next step in education. Considering I had always had an interest in aviation, I thought aeronautical engineering seemed a good fit. Through a series of events, I was offered a full-tuition scholarship at the College of Engineering at the University of Arizona (U of A). It was a great academic opportunity but also a chance to try to rebuild my life in a completely different environment.

My time at the U of A was an extraordinary experience. I lived with a loving couple who opened their home to me and had a positive influence on my life. Because the educational system of the US was unfamiliar to me, my hosts helped me choose courses for my first semester. One of the graduation requirements was to enroll in a humanities course for at least two years. Initially, I didn't understand why engineering students needed to spend time learning about humanities, but this course turned out to be the best one I took at the university. The course introduced me to the works of art, literature, and philosophy throughout human history.

My exposure to the richness of human thought and expression was a newly found treasure. The impact of this course was so great that I wanted to learn more about human beings, but I still loved science and engineering. Medicine seemed to blend these disciplines nicely, so I enrolled in the required pre-med science courses.

Wrestling with Apparent Conflict

One of the books we discussed in the humanities course was the Bible. This was the first time I had opened a Bible to read it critically. The course focused on the Bible as an ancient work of literature, and we spent time discussing the book of Genesis. Our teacher pointed out what she considered internal inconsistencies and highlighted the stark differences between Genesis and scientific findings. I had never read Genesis in detail. It puzzled me that this book seemed to be riddled with contradictions to what we know about the natural world. On a superficial reading, I thought that Genesis read like a myth. There seemed to be two creation accounts (chapter 1 and chapter 2), and the events of Genesis 1 just made no sense. It was easy for me to conclude that science had completely discredited these events—such as the presence of light before the creation of the Sun or the stars. I was shocked to think that billions of people believed in the God of the Bible and yet that Bible was full of mistakes in its first chapter. I didn't know what to make of it all, but the class moved on, and so did I.

An Influential Bible Teacher

While I was at the U of A, my mother started attending Bible study meetings in our hometown in Mexico. The first time I went back home I attended some of those meetings. The man who led the discussion had been a customs officer for the Mexican government. Though he wasn't highly educated, his knowledge of the Bible was excellent and he was a very good communicator. Up to that time I thought that only priests had the education and training to teach about the Bible, so I started meeting with him and asked questions about God. To my surprise, his answers were clear, coherent, and deeper than the answers that I had heard from the priests.

During one of our conversations, I asked him how he had decided to become a Bible teacher. He showed me his hands. He was missing both thumbs. He went on to tell me that he had been living in a world of lavishness, but also of bribes and corruption. One night while he was handling a shotgun, he accidentally fired it and blew off his thumbs. That event was a defining moment in his life. At that moment he realized that the path he was taking would eventually lead to more destruction—even to a premature death that would leave his young children fatherless. So, he sought God's forgiveness. He told me how he learned about the God of the Bible, how he finally believed that God had forgiven his sins, and how he decided to dedicate his life to teaching the Bible.

Those conversations made me reflect on my own life. I thought about the depravity of my actions in the past. It was hard for me to believe that if a good God existed, he would be willing to forgive people like me.

Pascal's Wager

Back at school in Arizona, I was introduced to Blaise Pascal and his book *Pensées*.[2] I was familiar with Pascal's important contributions to mathematics and physics, but the class focused on Pascal's works on theology and philosophy. At first, I thought that these were two different people. In my mind I couldn't comprehend that a mathematician would write about issues of faith.

I learned that Pascal had a dramatic conversion to Christianity. The events that led to his conversion are not known in detail, but he recorded his experience in the "Memorial," a note that was found in the lining of his coat after he died. His words "Certitude. Certitude. Feeling. Joy. Peace" indicated a sudden acknowledgement of both the reality of God as well as an emotional state filled with "Joy, joy, joy, tears of joy."[3] I was intrigued by Pascal's conversion.

During my childhood I was taught that miracle stories happened mostly in ancient times and that miracles were rare after Jesus died. Those few miracles that *did* happen in more recent history were unique. Therefore, the people who experienced these miracles were included in the Catholic pantheon of saints. Up to that point I thought that miracles, if real, would happen only to people who had a very special link to God. I was taught to never expect miracles and that the word was used in many mundane situations that had nothing to do with the supernatural. So, I grew skeptical of anybody claiming to have witnessed a miracle.

But here was Blaise Pascal, a "normal human being," not a saint, claiming to have witnessed what could be described as a supernatural event that touched his life in a very deep way. The event brought to my mind the conversion of Saul of Tarsus that I had heard about in the Catholic Church in Mexico. Acts 9 tells the story of Saul, a Jewish man hostile to Christianity who, while on his way from Jerusalem to Damascus, had a supernatural encounter with Jesus that transformed his life.[4] Pascal's conversion was similar, but he was a more contemporary man, one who had mastered mathematics and physics. He possessed a deep knowledge of the natural world and yet, after an unusual event, started arguing for the reality and importance of miracles in Christian belief.

When the class discussed Pascal's "wager," I took a close, sobering look at my life. The wager basically says that if a person believes in God and God does not exist, that person gains nothing. But if a person does not believe in God and God *does* exist, this person loses everything eternally.[5] What if God existed and I continued living as if he were not real? That would put me in a terrible predicament. I thought about the consequences and considered the possibility of God's existence, but I was still ambivalent. On the one hand I wanted to know if God was real, but on the other hand I was worried about a wrathful God who knew my history of sin.

Tension at Home

I was back in Mexico for summer break. My mother was now more involved with the Christian church she attended and I was able to ask the Bible teacher more questions. He explained some fundamental Christian beliefs and gave me a Bible in Spanish. During that time, I also started dating a young woman. She was a devout Catholic. My extended family had heard about my mother embracing the teachings from someone who was not a Catholic priest, and they warned that she was becoming a heretic. My dad sided with our extended family, which created a rift in my parents' marriage. When my girlfriend found out that I was involved in non-Catholic Bible studies, she distanced herself as I was getting more attached to her.

The tension within our family and with my girlfriend grew significantly over the summer. I returned to Arizona with a feeling of uneasiness. I tried to read the Spanish Bible, but it seemed too confusing. I thought that maybe opening it to random places would help, so I did that and read what was at hand.

One morning, my random reading led me to something about God not forgetting the sins of the wicked. I didn't like that passage. I turned to another random page. Again, it mentioned something about God punishing the wicked for their deeds. I reopened it a third time. Now the passage mentioned something about God punishing the wicked in haste. This set of three instances of punishing the wicked was a rather unusual coincidence, but I considered my past and a sense of fear rose within me.

My Accident and More Guilt

Later that day after a rainy morning, I was riding my motor scooter back from the university. Suddenly, as I tried to make a turn the scooter skidded, and I lost my balance. The scooter came to a stop under a pickup truck, and I tumbled a short distance down the road. I was not badly hurt, but the events from that morning and the passages in the Bible seized my mind. My fears were stoked. I wondered if getting "close" to God meant that my wickedness would be punished soon. I started trembling.

I was scheduled to take a bus trip to Mexico on the next day to visit my girlfriend, but I didn't want to go. My host family heard about my accident—though not my inner turmoil—but they weren't concerned. They convinced me to take the trip. Bruised and all, I ended up taking the bus. The bus broke down and it took me an extra 24 hours to get to my destination. When I finally saw my girlfriend and her family, it was understood that our relationship would not go any further because of my involvement with a non-Catholic group.

All these events in such a short period of time were deeply unsettling. The guilt over my past was now too much to bear. I didn't want to get close to a God who would certainly punish me forever. On the way back to Arizona, I raised my fists and spoke these words: "If you exist, get out of my life, and leave me alone. I want nothing to do with you."

God's Light Dispelled the Darkness

I decided to leave U of A. I moved back to Mexico City and went to medical school there to avoid the expense of medical school in the US as a foreign national. I also thought that my parents' guidance would be important at this point in my life since I always respected their advice. But when I moved back to Mexico City a month later, I found out that their marriage was on the brink.

This disturbing news contributed to my lapse into a deep and severe depression. There was no escaping my past, and if God was real, it looked like he was determined to destroy everything around me. I spent the next few days in bed thinking about how to end my life. I ate very little and didn't want to see anybody. My mother finally got me out of the bedroom so she could at least clean the room. Grudgingly, I made my way to the living room.

Sitting in the living room, I saw a book on a table. This wasn't a Christian book, but it had a few references to the Bible. I opened the book to a page that had Galatians 2:20 highlighted: "I have been crucified with Christ and I no longer live, but Christ lives in me. The life I now live in the body, I live by faith in the Son of God, who loved me and gave himself for me."

Suddenly, the room was filled with an unusually bright light and there was a presence in the room, although I couldn't attribute a material form to it. As I lifted my head, I felt as if a heavy weight was taken off my shoulders, and the severe depression disappeared immediately. I was suddenly filled with a joy that I had never experienced before.

I was stunned. How could this be possible? This event was not just emotional, it was physical as well. There was a peculiar light in the room and someone was there. I reached into my background in the sciences to try to make sense of what was going on. A light out of nowhere? The presence of an immaterial being in my living room? The sudden change from a state of severe depression to a state of incredible joy? The only words I could articulate were, "Who are you?" But there was no audible answer. I asked again, "Who are you?" But my words were met with silence. After a short while the presence left, but the joy remained. The whole event lasted mere moments, but I sat on the couch for a long time trying to make sense of what I had just witnessed and experienced.

Up to that point I wouldn't have considered such an event to be supernatural, but now the only reasonable explanation for the experience was that it was, in fact, supernatural. In my heart I had a new purpose: I needed to know who this being was. I thought it might be the God described in the Bible, but I needed to be sure. This was *not* the God that I thought I knew.

Worldview Testing
I'm thankful that God didn't reveal his full identity to me at that time. It's as though he gave me room to investigate and evaluate all possibilities. In my free time, I studied the different teachings of the main religions in the world and I learned the concept of a worldview.

Human beings, religious or not, live and behave according to a set of beliefs. We call this set of beliefs a worldview. Our motivations, our actions, and our decisions reflect that worldview. When I began evaluating the different religious beliefs, I found that many of these beliefs were either based on subjective claims or their claims were demonstrably false. The tools I used to evaluate these worldviews included the following:

1. Logical consistency: Are there any contradictions?

2. Empirical adequacy: Is there evidence to support the beliefs?
3. Experiential relevance: What is the impact of these beliefs on real life?

Logical Consistency
The test of logical consistency requires the use of critical thinking skills. Logic is fundamental to critical thinking, and the goal of logic is to reach truthful conclusions. A careful evaluation of the different worldviews can help us find inconsistencies in their beliefs. I found clear contradictions in some of the worldviews that I was evaluating; thus, I rejected them. I also found contradictions in some of the beliefs that I held, and I knew that I had to let them go, even if grudgingly.

Empirical Adequacy
The test of empirical adequacy was more complex. Some worldviews were easy to reject because they held historical or scientific claims that were demonstrably false. I was becoming more convinced that Christianity was the most coherent worldview, but the events depicted in the first chapter of Genesis remained problematic. It wasn't difficult to accept that the beginning of the universe required a cause outside of the universe, but the rest of the first chapter of Genesis didn't seem to fit with the timeline of what scientists know about the age of the universe and Earth. I tried to avoid this conflict for some time, but now I had to face it head-on.

I had spent my first few months as a Christian living in the grace of God. Living in that grace transformed my life. Grace is what makes Christianity stand out above all other religions of the world. There are no prerequisites to being accepted by Christ. He extends his love to whoever comes to him. He doesn't weigh the merits of our past. He cares about our future. It is because of his grace that I was given the privilege to spend a few moments in a room with him and receive his joy amid my depression.

But in medical school, even though I talked to my classmates about Christ and his grace, my evangelistic efforts were met with skepticism. Much of that skepticism stemmed from the apparent contradictions between the facts of science and the claims of the Bible. I was stumped. The speakers at my church didn't explain the details of the first chapter of Genesis during church meetings. And my friends who attended Bible school in Mexico City held to a young-earth interpretation of Genesis, which contradicted scientific findings. I realized that I had to research this topic on my own and try to resolve this apparent conflict. I believed that if God is a God of truth, then there must be an answer.

I would continue searching for the truth no matter where it led me.

Thankfully, I had access to one of the largest medical and scientific libraries in the country, so I routinely browsed through the scientific literature looking for articles that would address issues such as the origin of the universe, the origin of life, and the origin of humanity. I found dozens of articles in scientific journals that described the intricate complexity of the natural world from microbes to galaxies. It was not difficult to see the hand of a Designer in those discoveries. Yet, I struggled to come up with a model that would reconcile Genesis 1 with what I read in the scientific literature. Despite many months invested in this project, my models fell short one way or another. During that time if anybody asked me about the first chapter of Genesis, I had to admit that I didn't know how it fit together, but I had confidence that somehow it all made sense. The focus of my evangelism continued to be the grace of God. Still, the apparent disconnect between science and the Bible bothered me.

I moved back to the US in the summer of 1997 for my internship and residency training in internal medicine. The first year was demanding and I couldn't devote much time to anything other than clinical work. In the ensuing years I did have time, so I resumed the search for an answer on how to reconcile the book of Genesis and science. I browsed in bookstores for books on the subject, but these were scant.

The Science of God[6] by Gerald Schroeder was the first book I read on the subject because it was the *only* book at the bookstore that addressed the intersection of science and faith. It was insightful and provided an explanation for the apparent conflict posed by the first chapter of Genesis. Schroeder uses the whole universe as the frame of reference for the timeline of the events that occur on Earth. This makes time very pliable so it can accommodate both an old universe and creation events that happen over a period of six 24-hour days.[7] It was an ingenious explanation, but I wasn't convinced.

At that time, the Internet became an important tool for searching for just about anything, and during one of those searches I found an article by Hugh Ross in which he described his creation model approach for reconciling the Bible and science. I learned about his organization, Reasons to Believe (RTB), and visited the website. What I found was a very pleasant surprise. Many of the journal articles that I had collected were discussed in detail on the site and in RTB publications. The solution to the problems of the age of the universe, the age of the earth, and the events depicted in Genesis 1 was simple and elegant: The frame of reference for the timeline of biblical events was the surface of the earth and the days of Genesis could be interpreted as long but finite periods of time without compromising the integrity of the text. I felt as if I had hit the jackpot! This realization fulfilled the last test of empirical adequacy that I was looking for.

Experiential Relevance

The test for experiential relevance is a more subtle one. It is not uncommon to hear about people who have had unusual experiences. I remember crossing paths with some of them during my medical school years in the psychiatric wards or in the emergency department after a wild drug party. An unusual experience can be the product of physical, chemical, or psychological disturbances in an individual or even just a dishonest attempt to fool the naïve. Modern medicine has many tools for evaluating physical and chemical disturbances. A computed tomography scan of the brain, for example, can discover a brain tumor. A urine test can find substances that can alter one's perception of reality.

Psychological disturbances and dishonest claims require a more in-depth evaluation. It's important to remember that subjective experiences on their own cannot be falsified, but when those experiences become part of a person's worldview, that worldview can and should be evaluated.

Lifelong Testing

Evaluation of Christianity's claims should be a lifetime task for Christians. We are called to "test all things. Hold on to what is good" (1 Thessalonians 5:21, CSB). We will continue hearing about religious leaders, authors, scientists, and philosophers who claim to show that Christianity is not true or who give alternative (nonbiblical) views on Christianity. As a Christian who takes the Word of God seriously, I have continued to explore the relationship between science and the Bible throughout my professional life. I've found that as we discover more about the natural world, the case for a Creator becomes stronger and stronger. I don't want to let complacency get in the way of my search for truth.

On the day when God met me in my living room, I knew that he was real. Even though it was a very powerful experience, it was still a subjective event. My God-given capacity to reason and my knowledge of the natural world provided me with the compelling evidence to conclude that the person who met me was, in fact, the God of the Bible. To this day I still ask the question, "Why me?" But as Paul says in Romans 1:20, "For since the creation of the world God's invisible qualities—his eternal power and divine nature—have been clearly seen being understood from what has been made, so that people are without excuse."

Nobody needs a special visit from God to know that he is real. We only need to lift our eyes to the sky, take a stroll on any hiking path, or look under a microscope to see God's glory. Blaise Pascal understood and felt the power of God's natural revelation. I suspect that he kept his memorial in a very intimate, almost secret place so that he would not place a subjective experience as the core of his arguments for the existence of God. Paul's dramatic conversion provided the people of his time with evidence of how God can transform lives. Both men used the power of reason and the evidence found in the natural world to reach out to non-Christians. I hope to do the same in my lifetime.

Chapter 4

An Obstetrician's Reflections on Life and Suffering

Christina A. Cirucci

As I looked over the Rohingya refugee camp in Bangladesh, I was distraught. Nearly a million people lived there, having fled from torture in Myanmar. As a physician, I had seen plenty of suffering, particularly on other working trips in Bangladesh, but this misery was unimaginable. The suffering of one person is heartbreaking, but here were nearly a million suffering people in this camp. Many had lost loved ones and had been mutilated, tortured, and raped.

At the time, I was working a few hours away at a mission hospital. These refugees frequently sought medical care at our hospital. I saw a young boy with severe burns. His mother had perished in the same fire that had burned his body. People with countless other calamities came to the hospital every day. Even before the Rohingya crisis, the suffering I saw on my many visits to this poverty-stricken country was inconceivable.

As an obstetrician/gynecologist, most of what I saw was childbirth related. I saw mothers who had hemorrhaged and died in labor, and babies who never had a chance. I saw women who had suffered in obstructed labor for days. Not only had they ended up with a dead baby, but they had become outcasts in their communities from urinary incontinence due to a fistula.[1] As an obstetrician, I've seen both ends of the spectrum of life: the joy of childbirth and the sorrow of death. Many experiences along that spectrum have presented intellectual and emotional challenges to my faith in the God of the Bible. To be a physician and maintain my Christian faith, I've had to reconcile my faith with science. I've also had to address the dilemma of human suffering—including my own— in relation to an all-good and all-powerful God.

Beginnings: My Story

I chose to live my life for Jesus Christ when I was in eighth grade. That decision defined my life going forward. Jesus said that following him meant denying myself and valuing him more than anything else in my life (Matthew 16:24). Jesus described the kingdom of heaven as a pearl of such great value that the pearl merchant sold all he had to buy it (Matthew 13:45–46). Following Christ is more than agreeing with particular doctrines. It's a complete surrender of my life and a commitment to live my life for him.

Intellectual belief is a necessary foundation, though, and I begin my story there. I was raised in a nominally Christian home, attended church every week, and identified as a Christian. I believed in Jesus and thought this intellectual assent made me a Christian. When I was eleven, I started reading the Bible every day, a discipline that would change my life. I was surprised to learn that there was much more to being a Christian than I had realized. If the Bible was true, then Jesus Christ changed people's lives. I longed for this difference in my life and prayed every day, "God, please show me how to be a true Christian."

When I was thirteen years old, I attended a Christian camp where I came to understand what Jesus's death on the cross meant and that I had a choice to make. I clearly remember that night, in a cabin in the woods, surrendering my life to Jesus Christ. I received his forgiveness for my sins and chose to follow him. Although my conversion experience was not externally dramatic, it defined my life going forward. There was an evident change in my heart that has continued to this day, decades later. It all began with reading the Bible and absorbing that knowledge for myself.

After high school, I earned a degree in mechanical engineering and worked in the field for seven years. As an engineer, I didn't have any problem reconciling my Christian faith with my vocation as I encountered few ethical issues. However, I wasn't too fond of engineering, so I decided to pursue the vocation I longed to do. I went to medical school and then completed a residency in obstetrics and gynecology (OB/GYN).

I love working in medicine, but it has brought many personal and ethical challenges. Medicine is not just a science but an art. People's lives are on the line every day. As a physician, I've been faced with life and death in countless circumstances: when someone was dying before my eyes, when seconds mattered in getting a baby delivered to save her life, or when I had to make a treatment decision even though there was no clear-cut good choice.

Ethical Issues at the Beginning of Life

I've encountered a plethora of ethical dilemmas as an OB/GYN physician that have challenged my Christian faith. In medical school, another Christian student told me that he had eliminated OB/GYN as a career option because of the many ethical issues. These issues made me tremble, but I didn't think I should shy away from hard things as a Christian. If what I believe is true, was not my God big enough to provide clarity?

One of the challenges in OB/GYN is dealing with early embryonic development and whether to treat the embryo as a person deserving the right to live. The Bible indicates that an unborn human is a person from conception. Psalm 139:13 states, "For you created my inmost being; you knit me together in my mother's womb." Also, "My frame was not hidden from you when I was made in the secret place, when I was woven together in the depths of the earth. Your eyes saw my unformed body; all the days ordained for me were written in your book before one of them came to be" (Psalm 139:15–16).

As a Christian, the view that life begins at conception impacted many aspects of my medical training and clinical practice. It meant that I had to thoroughly research the mechanism of action of each type of contraception and decide whether I could prescribe it in good conscience. It meant that I did not perform abortions and had to make sure I worked for an institution that did not expect that of me. It meant that I had to communicate my stance gently and respectfully with coworkers and make sure that we could, at a minimum, agree to disagree. Many of my fellow residents and coworkers didn't struggle with considerations of when life begins. It seemed like an easier path; nevertheless, I knew that my commitment to Christ required that I heed my conscience.

To take this kind of stance, I needed to be sure that my faith was backed by science. Long before I learned how to deliver babies, I studied the astounding process by which they were formed. A human being has 23 pairs of chromosomes (diploid, 2N). The oocyte (egg) and the spermatozoon (sperm) each have 23 single chromosomes (haploid, 1N). The spermatozoon penetrates the oocyte's zona pellucida (wall) at fertilization, and the two cell membranes fuse. The DNA of both the sperm and the egg duplicate. Then the DNA intermingles and divides, resulting in a unique organism with 23 pairs of chromosomes (diploid, 2N).

This one-celled organism has the exact genetic makeup it will have his or her entire life. The human's sex, eye color, and other factors are determined at fertilization. At the end of the second week, the embryo has a primitive uteroplacental circulation,[2] and at the end of the fourth week, there's a beating heart.[3]

I have seen that tiny heartbeat countless times on ultrasounds just four weeks after conception.[4] The study of embryological development confirmed what the Bible says: human life begins at conception.

In the past, abortion advocates denied that a preborn baby is a human. But now, as Nancy Pearcey states in her book *Love Thy Body*, "virtually all professional bioethicists agree that life begins at conception."[5] Due to advances in scientific understanding, the abortion argument is no longer whether the preborn baby is *life* but whether it is a *person* deserving the same rights as those who live outside the womb. I find that argument untenable. If personhood does not begin at conception, then when does it begin? Typically, babies are afforded the rights of personhood once they are born (although even this demarcation has been challenged). Is the baby then a person *after* delivery but not five minutes *before*? Is the baby a person only when she is outside the womb? Most babies born at 25 weeks survive, and some survive at 22 weeks.[6] If the preborn baby is not a person until he can live outside the womb, then does that mean that an unborn baby at 22 weeks is a person, but at 21 weeks, 6 days is not a person deserving of the rights the rest of us enjoy? Does personhood suddenly emerge in that final hour?

I don't find this reasoning to be a rational way to draw the line for when human life and personhood begin. As Pearcey states, "Clearly, it would take a dramatic transformation to turn a mere human organism with no rights into a person with an inviolable right to life. But there is no scientific evidence of such a transformation—no single, dramatic turning point can be empirically detected."[7] Furthermore, if personhood emerges sometime later in embryonic or fetal development, we have no way of knowing when that is and must err on the side of not killing a human being.

The fact is that there is no specific point of transition. At fertilization, this tiny organism is a human person. The Bible supports the claim that a preborn baby is a person—no matter at what stage of development—and what I see every day as an obstetrician confirms this truth (Psalm 139:13–16; Jeremiah 1:5; Luke 1:41, 44). Both my scientific training and my faith convince me that this beautiful life, with all its genetic materials at conception and a beating heart a few weeks later, is not just tissue. It is the crown jewel of creation formed by the Creator and Designer of the universe.

The Problem of Suffering and Loss
It is a joy to deliver babies and bring new life into the world. Most of the situations that I see as an OB/GYN physician are happy ones, but I've witnessed

many tragic situations too. Some of them are particularly heartrending: still-born babies, complications at delivery, maternal death, and gynecologic cancers. The joy I experience from delivering a screaming baby to ecstatic parents in one room can be muted by the knowledge that my patient in the next room is suffering a terrible loss. Moreover, outside my small sphere of life, there's exponentially more tragedy and suffering: the refugee camp in Bangladesh, the young man paralyzed in a tragic accident, the teenage girl trafficked into slavery.

Many of us struggle with the issue of pain and suffering. Why do people suffer? Why do people die? Where is God in all this pain? Realizing the magnitude of suffering causes many people to question how a good God can allow such evil. It leads some to argue that if God exists, then he is either good or powerful, but he cannot be both. Philosophers, theologians, atheists, and agnostics have wrestled with this question for centuries. I certainly won't be able to answer the question here, but I want to share my journey and the small insight I've gained through it.

My Suffering and Loss

My own story of hardship involves cancer, the dreaded disease I did everything to avoid. No one wants to get any cancer, but as an OB/GYN doctor, the one cancer I did *not* want to get was ovarian cancer. I had seen many patients suffer and die from this terrible disease. Ovarian cancer was a death sentence. Unlike some cancers, there is no screening method or early detection for ovarian cancer. It is typically found in a late-stage and is not curable.

Women with a BRCA1 gene have a 44% risk of developing ovarian cancer and a 72% risk of breast cancer.[8] Imagine my shock when I learned that I carry this gene. For women with BRCA1, cancer can sometimes be prevented by removing the ovaries. The day I found out that I carried the BRCA1 gene, I called my colleague, a gynecologic oncologist, and asked her to remove my ovaries surgically. I planned to do anything I could to prevent ovarian cancer. Within two weeks, I underwent the surgery. Everything appeared normal at the surgery, and I planned to move on with my life. A few days after the surgery, my doctor dropped the bombshell when she called and told me I had cancer. I had done everything I could as fast as possible to prevent cancer. Yet I still contracted cancer. Not just cancer. Ovarian cancer.

After another surgery and chemotherapy, I returned to work and hoped my life would return to normal. My doctors were optimistic that since the cancer had been detected early, there was a possibility of a cure. Unfortunately, cancer recurred in my abdomen and chest cavity a few years later. It was a whole new scenario. My doctors no longer gave me hope for a cure. It no longer mattered that my cancer was initially found early because now it had recurred. I was given three to five years to live.

When ovarian cancer recurs once, as it has for me, it will recur again and again, with successively shorter intervals, until it leads to death. Although I am in remission now, any day my cancer will rear its ugly head and take over my body. As a gynecologist, I knew the picture—and it was not good. Not only do I have cancer hanging over my head, but the treatments and complications from cancer have caused further medical problems and side effects. I'm living a whole new normal due to the aftereffects and complications of cancer and its treatment. I was passionate about my work as a physician, but now I'm not physically able to do this work. I treasured the opportunities to volunteer in Bangladesh and I built many relationships there, but I haven't been able to return since my cancer recurrence. Simple daily tasks are often a challenge. Dealing with my cancer and all its accompanying losses has been a trial.

However, my level of suffering is not even close to that of so many others. I live in a free country, have food on my table, and relative peace in my environment. I can't compare my cancer to the suffering of people who've been trafficked, tortured, or imprisoned. But cancer has undoubtedly rocked my world. It's easy to question God. Is my faith a hoax? Is God there? How can a loving God permit such things? Does God even care? Although I continue to struggle, I've concluded that—cancer or not—God is there, and he is the powerful and loving God of the Bible.

Reflections on the Meaning of Suffering

Simply stated, either God exists or he doesn't. Suffering doesn't change that. I don't know why God allows suffering. We humans will continue to ponder this issue as long as we live on the earth. The problem of pain and suffering exists not only for Christians but for people of all faiths, as well as atheists. My inability to grasp how suffering and God fit together does not mean that God does not exist. Scripture doesn't give easy answers, but it does provide wisdom and insight. Just as the Bible led me to faith as an adolescent, it has directed me throughout my life and provided guidance in my struggle.

The book of Job describes the enormous suffering that God allowed in Job's life and how Job stayed faithful to God. Although God restored many of Job's losses, he never gave Job a reason for his suffering. Nor did he apologize. God's answer to Job was not, "So sorry about this, Job," but rather a powerful exposition of who he is: the God who laid the earth's foundation and ordered the stars, the sea, and the clouds. God asked Job, "Have you journeyed to the springs of the sea or walked in the recesses of the deep? Have the gates of death been shown to you? Have you seen the gates of the deepest darkness? Have you comprehended the vast expanses of the earth? Tell me if you know all this" (Job 38:16–18).

In *my* suffering, I want to know why, and I want relief. God has not provided either of these for me, but he has walked the path with me. One of the lessons I'm learning from my cancer journey is that this life is not supposed to be easy. Many of us simply want a good family, a stable job, and a vacation now and then. That doesn't seem too much to ask, and although there's nothing wrong with these blessings, God does not promise them. The Bible doesn't teach that life on this earth will be comfortable. In fact, Jesus said we would have trouble (John 16:33). Like most people, I don't want to suffer. I want God to make my life comfortable. C. S. Lewis said, "We want, in fact, not so much a Father in Heaven as a grandfather in heaven—a senile benevolence who, as they say, 'liked to see young people enjoying themselves,' and whose plan for the universe was simply that it might be truly said at the end of each day, 'a good time was had by all.'"[9]

I'm disturbed by prosperity preachers who say that God will give you a good life if you just have faith. That's not true, at least not how they say it. God will give us a "good life," but it may not be on this earth, and it likely will not be the way we perceive it should be. To those who follow him, God will give more blessings than a family, a job, and financial security, though not necessarily in this life. The apostle Peter tells us to not be surprised by suffering (1 Peter 4:12). And the apostle Paul wanted "to know Christ—yes, to know the power of his resurrection and participation in his sufferings" (Philippians 3:10). Unfortunately, Christianity is sometimes falsely presented as a way to a prosperous, carefree life, but that's not the path that Jesus offers.

The Bible gives some perspective on the brevity of our life on Earth. The psalmist said that even if we live to be 80, the best of our days is but trouble and sorrow, and then they quickly pass (Psalm 90:10). When I ponder the age of the earth (4.5 billion years old) or that of the universe (13.8 billion years old), what significance is my life even if I live to 100? The Bible says that my life is like a mist that appears for a while and then vanishes (James 4:14). When I ponder the size of the universe and the distance of the stars, it's easy to feel insignificant.

However, the Bible doesn't say that humanity is insignificant and worthless.

Instead, Scripture speaks much about human worth and God's care for us. The Bible teaches that I was significant even when I was in my mother's womb (Psalm 139). When I consider how small I am in the context of God and the universe, I better understand the vastness of this world, the greatness of my God, and the significance of eternity. This perspective provides a context for my suffering. Maybe my cancer is not quite the tragedy that it often feels like in my small perspective.

God has not cured my cancer or even given me a reason for it. I've lost my job, health, and the ability to do many things I enjoy. I can get bitter about my cancer (and, at times, I do), but that misses the whole point. This life is not about me. God has shown me that there's a bigger picture. Although we understandably desire comfort and happiness in this life, that's not the meaning of life. There's something more significant going on. In *The Problem of Pain*, Lewis eloquently states, "Man is not the centre. God does not exist for the sake of man. Man does not exist for his own sake . . . We were made not primarily that we may love God (though we were made for that too) but that God may love us, that we may become objects in which the Divine love may rest 'well pleased.'"[10]

One glimpse into the bigger story that may give meaning to suffering comes to us in the story of Joseph in the book of Genesis. After Joseph's jealous brothers sold him into slavery, Joseph was deported to Egypt, where he was falsely accused of unwanted sexual advances and spent years in prison as an innocent man. Many years later, Joseph encountered his brothers and said to them, "You intended to harm me, but God intended it for good to accomplish what is now being done, the saving of many lives" (Genesis 50:20). God used Joseph's trials to protect the Hebrews during a big famine and to grow the nation in Egypt.

Unlike Joseph, most of us will not see the big picture of how everything works out according to God's plan in this life. Sure, I've had opportunities, times with people, and different chances for interaction that would never have come about if I had not gotten cancer. In my perspective, though, these blessings do not outweigh the trials. Honestly, I would rather have my old life back. The good things that have resulted from my cancer provide only a small consolation because I've lost so much. The Bible says, though, that we only see in part right now (1 Corinthians 13:12).

An old poem called "The Weaver" describes each life as a tapestry in which we only see the underside, not the beautiful result on the front.[11] It's a helpful analogy to demonstrate the perspective of the bigger picture, but I prefer to think that, in this analogy, my life is not the whole tapestry but just a thread within it. Not only do I *not* get to see the finished tapestry right now, but my life is not the tapestry. There's a larger domain. Those things that seem essential may not be. Does the president of the country or the winner of the Nobel Prize look back on his life and wonder why, at three years old, he chose a yellow lollipop instead of a red one?

Many circumstances that seem so significant now may play a different role in light of eternity. Likewise, I believe that when I die and am in the presence of God for all eternity, I will look back on my cancer and realize how small that problem was. There is meaning to suffering beyond what I can see during my earthly life. I don't minimize the many struggles of this world, but the sufferings of this life are only a speck in light of eternity and they bear a significance far beyond this mortal life. The Bible says that our troubles are achieving an eternal glory that far outweighs them all (2 Corinthians 4:17).

God does not promise the things we might think we need—like answers to "why" questions—but to those who believe and trust in him, he promises his presence, eternal life, peace, sustenance, and much more (Deuteronomy 31:6; Joshua 1:9; John 3:15, 4:14, 5:24, 14:27, 16:33; Psalm 55:22; Psalm 54:4). No matter what happens in this life, God is still with me. As the psalmist says: "God is our refuge and strength, an ever-present help in trouble. Therefore, we will not fear, though the earth gives way and the mountains fall into the heart of the sea" (Psalm 46:1–2). Maybe the earth will give way, and the mountains will fall into the sea. Maybe my cancer will take my life later this year. Maybe another pandemic will happen. No matter what, God is there, and he has a plan far beyond what I can imagine.

Staking My Life on the Truths of Scripture

These ponderings do not explain the refugee crisis in Bangladesh, the stillborn birth, or my cancer, but I know that suffering does not change the facts about God. If there is no God, then suffering is still a dilemma. I believe there is a bigger purpose, a picture in the tapestry that I cannot see from my perspective as one thread. It's a daily struggle for me to maintain this attitude, but I know God is there, and my difficulties do not change that. If Jesus Christ is God and died for me, then my suffering is not for nothing. It's part of a bigger plan that I can't see. Furthermore, I don't walk alone in my suffering—God is with me in it. I may never understand it and I certainly will never like it. Even so, there is a God who loves me in it.

I've studied the Bible and staked my life on its truths. It's consistent with the truth I know in medicine and see in nature. I've seen the miracle of birth, and I believe God is the Creator of life. I've walked through suffering, and God has been with me every step of the way. Like Job, I think it would be nice if God would give me some answers. However, I realize that if he is truly God, then the answers must be beyond my understanding. Life with or without God is not an easy path, but for those who walk that path with the God of the universe, he gives meaning. The same God who formed the baby in the womb is the same God who walks with me in trials. Life with God is a pearl of great price, and I have sold my life to have it.

Chapter 5

How Could a Jew Believe in Jesus?

David L. Block

I grew up as an orthodox Jewish boy in Krugersdorp, South Africa. I regularly attended *cheder*, a traditional elementary school teaching the basics of Judaism and the Hebrew language. This learning prepared me for the Torah and Haftarah readings on the day of my Bar mitzvah. My oral readings in Hebrew from the Torah at our *shul* (synagogue) took place on Shabbat, March 4, 1967. I was 13.

I was enthralled, sitting in shul, as our learned rabbi expounded how God was a *personal* God. He would speak to Moses, to Abraham, to Isaac, to Jacob, and to many others. I pondered how I fit into all of this.

Spiritual Restlessness
By the time I entered university, I became deeply concerned that I had no assurance that God was indeed a personal God. Where was the personality and the vibrancy of a God who could speak to me?

At the University of the Witwatersrand, Johannesburg, I began studying for my bachelor of science degree in applied mathematics and in computer science. I also took an interest in astronomy and was elected a Fellow of the Royal Astronomical Society of London at the age of 19. As a student, I became friendly with Lewis Hurst, then a professor of genetics and of medicine. He also had a great interest in astronomy. For many hours we would discuss the complexities of the cosmos at his home in Krugersdorp. I delighted in explaining to him fundamentals in astronomy such as "black holes" and "quasars."

Intellectually, I was satisfied. The elegance of the mathematical formulation of general relativity fascinated me. My first research paper on that theme was submitted to the Royal Astronomical Society of London in 1973 and published by that society one year later.[1] Nevertheless, spiritual questions—deeply personal spiritual questions for which I had no answers—haunted me. Inwardly, something, or *Someone*, was missing.

My friendship with Professor Hurst grew over the next few years and I continued sharing my thoughts and feelings about the cosmos with him. "The universe is so beautiful," I exclaimed, "from both visual and mathematical perspectives." The professor listened intently. Was the incomprehensible-yet-comprehensible world which I studied, indeed one of purpose? And who was behind that purpose?

To be brutally honest, I did not *know* God. What concerned me, deeply so, is that the universe is so large, so immense. Is physical reality the sum total of our existence? I often reflected on this question as a young university student in South Africa.

Hurst told me that there were answers to all the questions I was asking. He knew that I came from an orthodox Jewish family and asked if I'd be willing to meet with a friend of his, the Reverend John Spyker. My Jewish parents had taught me to seek answers where they may be found—and so I consented to meet with this Christian minister.

Identifying My Stumbling Stone

The voice of the Reverend John Spyker commanded attention, for he spoke with authority. He took his Bible in his hands and turned the pages to one verse in the book of Romans in the New Testament. In that letter the apostle Paul affirms that Y'shua (Jesus) is a stumbling stone to my Jewish people, but also that those who freely choose to believe in Y'shua shall never be ashamed.

"As it is written, Behold, I lay in Sion a stumblingstone and rock of offence and whosoever believeth on him shall not be ashamed" (Romans 9:33, KJV). The words read by John Spyker deeply penetrated my heart.

I think there's no work on canvas that better describes *the* biblical stumbling stone as the masterful painting entitled *Obsession* (1943) by the Jewish artist Marc Chagall (see figure 5.1). A house—possibly a house in Vitebsk—is burning. A horse, harnessed to a cart, stands frozen with fear. A mother attempts to move the horse forward. But the horse will not move. Cornelia and Irving Sussman describe the painting in these words: "Almost filling the center of the canvas is a burning house. . . . No path of escape is open. The fallen Christ blocks the path. The path is blocked; there is no way out, no hope; for there can be no hope unless the cross stands upright."[2]

Figure 5.1: *Obsession* **by Marc Chagall (1943)**
A house is on fire. A horse, harnessed to a cart, refuses to move. Fear looms large. Chagall's rooster, his bird of peace, lies dying. The fallen Christ lies at the door. Would I too, stumble over that stumbling stone placed in Zion—the Jewish Jesus? *Credit: AAl Fotostock.*

I was spiritually encased in a house, burning not with fire but with fear. Was life an unspeakable tragedy? Would my grave speak the final word? Would my candle be forever extinguished? Was my pain, His pain? Would I, too, stumble over that Stone placed in Zion, the King of the Jews?

By an infusion of divine grace, the truths of Romans 9:33 blazed in my heart. Everything suddenly (instantaneously) became very clear to me. Y'shua (Jesus) had been *my* stumbling stone. Jesus had fulfilled all the messianic prophecies in the Hebrew Scriptures—such as where the Jewish Messiah would be born (Micah 5:2), how He was to die (Psalm 22:8; Isaiah 50:6), and much besides.

While many of my Jewish people were still awaiting the Messiah, I suddenly knew that Jesus is the living Messiah! All I had to do was respond to his grace. I immediately asked the reverend to pray for me. I surrendered my heart and my reason to Jesus that day. The Spirit of Jesus infused every cell of my being. That was in October of 1976. I was 22.

What About the Holocaust?

It might seem strange to some that a young scientist and orthodox Jew should come to faith in Christ. It's been several decades, and I'm asked a recurring question: How can a Jew believe in Jesus in light of the Holocaust?

Here's how I've responded. The *representations* of Jesus given to the Jewish people over the centuries have often been nauseating, in the extreme. The Holocaust is uppermost in many Jewish minds. But it's not only the Holocaust. Any mention of Jesus to Jews often invokes memories of the Inquisition, too, with its methods of torture, murder, and forced baptisms. The Jewish people have been despised throughout history. Professor Jakób Jocz (1906–1983), a leader in the international Hebrew Christian movement, summed it up this way: "The Jew, who gave the world the prophets and apostles, the Bible, and, humanly speaking, our Lord Himself, became the most despised figure amongst men."[3]

Literary giant in Hebrew literature Yosef Haim Brenner (1881–1921) paints the New Testament with a fully Jewish brush, asserting that the New Testament is "*our* book, bone of *our* bone and flesh of *our* flesh"[4] (emphasis added). Had I indeed not completely deserted Judaism in 1976 and moved over to the other (Christian) camp?

Of utmost importance here is the need to understand the radical difference between *religion* (understood as simply keeping long-held traditions) and a deeply abiding *relationship* with God. This kind of religion may have nothing whatsoever to do with God and with the fruits of God's Spirit:

> But the fruit of the Spirit is love, joy, peace, forbearance, kindness, goodness, faithfulness, gentleness and self-control. Against such things there is no law (Galatians 5:22–23).

How does one reconcile these fruits with some of the most monstrous deeds ever conducted by mankind, many under the mantle of religion?

Visit to the Death Camps

It was in the year 2017 that I visited Auschwitz-Birkenau in Poland, having been invited to address a conference near Krakow. The rotten fruits displayed at Auschwitz-Birkenau belong to a universe characterized by an abyss of utter darkness and destruction. I sat down on the railway line at Auschwitz. Silence everywhere. Not even the birds dared sing.

What did Hitler and the Nazis come to do with their "final solution"? They came to steal, kill, and destroy. In that painful moment I remembered the words of Jesus in John 10:10: "The thief comes only to steal and kill and destroy; I have come that they may have life and have it to the full." But how could I follow Jesus knowing that some Christians allowed the Holocaust to happen?

First, Jesus revealed himself to me. He manifested himself to me at the door—the door of my heart. I knew that He was on my way, not in my way. Before my conversion there were two distinct universes, one of utter darkness—humanity's condition of sin without Christ's forgiveness—and one of glowing light. By God's grace I entered the light.

Second, Jesus is the fulfillment of all Messianic prophecies as written in our Jewish Scriptures, such as in the book of Isaiah (Isaiah 61:1; see also Luke 4:21).

Third, as far as the east is from the west, so is the love of Jesus separated by a gulf from the Nazi universe of terror.

Another Chagall masterpiece, *The White Crucifixion* (figure 5.2), has helped me to think through what it means to embrace Christ as a Jew. Why the ladder in that figure? Sussman and Sussman elaborate:

> Is it not possible that Chagall may yet be thought of as one who has set up a ladder against the Cross for many to climb? Is it not possible that he will help many to see the ladder which leads from Jewish suffering to Christ's, from all men's sorrow to His, and not only to His Passion, but also to His peace? Yes, there is a ladder from the earth of even modern tears to the hill of peace, and Chagall seems its painter.[5]

On a personal level, not only was Jesus crucified about 2,000 years ago, but I continue to crucify Jesus, moment by moment, in thought, word, and deed, whenever I sin (2 Corinthians 5:21; 1 John 1:5–10; Galatians 2:20). The good news is that my pain becomes His pain at the Cross (Isaiah 53:10, Romans 5:15). My forsakenness and anguish become His forsakenness and anguish (Psalm 22). He carries all my shame at the Cross (Hebrews 12:2). If anyone knows the horrors of Jewish pain through the centuries, it is the luminous Figure of the Nazarene.

What About the Inquisition?

Scholars have wrestled with the problem of evil for centuries. Reflecting on the Inquisition, Russian novelist Fyodor Dostoevsky perceptively distinguishes between the exercise of power, control, torture, and death (as practiced under

Figure 5.2: *The White Crucifixion* by Marc Chagall (1938)
The artist emphasizes the Jewish identity of Jesus in several unique ways: he replaces his traditional loincloth with a prayer shawl (*tallit*), his crown of thorns with a head cloth, and the mourning angels which customarily surround Jesus are replaced by three biblical patriarchs and a matriarch. The four are clad in traditional Jewish garments. *Credit: AAI Fotostock.*

the banner of religion during those horrid times), and Jesus. In "The Grand Inquisitor," the short story in the novel *The Brothers Karamazov* by Dostoevsky, the author presents Jesus to us in human form on Earth (as in John 1). But He walks not in Galilee, but in Spain—during the Spanish Inquisition.[6] The Grand Inquisitor meets Jesus face to face. His reaction? The Grand Inquisitor says to Jesus, "I shall burn Thee for coming to hinder us. For if anyone has ever deserved our fires, it is Thou. To-morrow I shall burn Thee. *Dixi.*"[7]

Jesus himself is not exempted from being killed—by fire—at one of those nauseating auto-da-fés (public ceremony and execution) in Spain. That had been decided by the Grand Inquisitor. What I take from this great novel is that Jesus bore the punishment for all our inhumanity to one another.

Figure 5.3: Cosmic Dust Clouds
Imaging of the region surrounding the reflection nebula Messier 78, north of Orion's belt, shows dark clouds of cosmic dust threaded through the nebula like a string of pearls. *Credit: ESO/APEX(MPIfR/ESO/OSO)/T. Stanke et al./Igor Chekalin/Digitized Sky Survey 2.*

With my Jewish background, I cautiously distinguish between actions conducted by sword and by torture in the name of religion—such as the Inquisitions—and actions by those who obey the will of our Lord and who follow His voice of grace (see Matthew 7:22–23). Christians have not always been faithful to follow that voice, but many have and continue to do so.

Research Guided by the Creator
At the start of my career, I knew that it was God who had been inextricably involved in my research. (Several decades later, that is still true.[8]) I studied the shapes (morphologies) of galaxies, with their dark and pervasive clouds of dust. And when I read my King James Bible, God highlighted one specific verse

Figure 5.4: The Pavo Group of Galaxies
The Pavo group of galaxies photographed at the prime-focus of the Víctor M. Blanco Telescope (a 4-metre aperture telescope located at the Cerro Tololo Inter-American Observatory) in Chile. We are looking almost 300 million years back in time. The galaxy at center (NGC 6872), measures nearly 750,000 light-years from tip to tip. The author identified this specific galaxy as one of the largest known spiral galaxies in the universe. *Credit and copyright: Association of Universities for Research in Astronomy Inc. (AURA); all rights reserved.*

to me (Isaiah 45:3): "And I will give thee the treasures of darkness, and hidden riches of secret places, that thou mayest know that I, the Lord, which call thee by thy name, am the God of Israel."

It was God speaking *to me*, through the Scriptures.[9] A revelation from the heart of God, by means of His Holy Spirit.

What are these treasures of darkness, to which I would devote a lifetime of research? They are clouds of cosmic dust grains (figure 5.3)—the stuff of which you and I are made (Genesis 1:27). Such clouds may partially or totally obscure remote starlight, but infrared cameras can freely "penetrate" such dust clouds.

From where do these dust grains originate? The starting point for producing interstellar dust is in the atmospheres of cool, older stars. Minute rock-like

Figure 5.5: Infrared Cameras Penetrate Dust Masks of Galaxies
NGC 309 presents an exquisite (symmetric) two-armed spiral morphology in the infra-red. Ninety-five percent of the total luminous mass in NGC 309 lies not in the disk of young stars and gas seen in figure 5.3, but rather in the dust-penetrated image above, secured at the Mauna Kea Observatories in Hawaii. *Credit: D. L. Block, R. J. Wainscoat & Nature.*

particles called silicates are blown from the atmospheres of these stars into surrounding space. These "submicron" particles (smaller than one micron) are blown away from their parent stars, and then they drop down in temperature to the coldness of space—temperatures as low as minus 253°C.

The degree of excitement and joy I experience in studying the *City of God* above our heads cannot be described. In 1979, I identified one of the largest known spiral galaxies[10] in the universe (figure 5.4). Its spatial extent spans some seven Milky Way Galaxy diameters, or about 750,000 light-years from tip to tip.

Over the years, I've also had access to some of the most advanced infrared cameras ever developed. In 1990, my research group studied the galaxy NGC

309 through infrared eyes at the Mauna Kea Observatories in Hawaii. Could it be true that, in the eyes of the infrared, NGC 309 was hinting at a fundamental new and general "hidden symmetry" in nature? Could it be that behind the dust shrouds of many spiral galaxies there lies a beautifully symmetric stellar backbone?[11] The answer was a resounding yes! (See figure 5.5.)

It was as if a celestial musician was using an instrument with bow and strings, with the gargantuan bow being the "dust-penetrated" structure interacting with the young stars and those pervasive gas and dust clouds seen in optical photographs.

God's favor was upon our research. His promise to reveal the hidden treasures of darkness (Isaiah 45:3) was unfolding, moment by moment. We submitted our research on the dust-penetrated image of NGC 309 to the journal *Nature*. (It was in *Nature* that Watson and Crick unlocked the DNA code.) Not only did *Nature* publish our discovery, but they also featured our infrared image on their cover.[12]

Other exciting discoveries lay in store. For example, a team of my collaborators established that astronomers had not detected 90% of the dust mass in our universe.[13] An analogy may be helpful. Imagine you are about to watch ballet dancers on stage. Curtains lie between the dancers and the audience. What we had discovered was that, instead of simply one thin curtain obscuring the ballet dancers, there were nine other such curtains. In this analogy, the mass of the curtains was to increase by 90%. No wonder cold cosmic dust has become such a central stage in astronomy and in astrophysics!

The Spitzer Space Telescope, in orbit about the Sun, also opened our eyes, yet further, to the infrared universe. Onboard is an infrared camera designed by a very close friend and collaborator at the Harvard-Smithsonian Center for Astrophysics, Giovanni Fazio. When that telescope was pointed at the Andromeda Spiral Galaxy, some 2.2 million light-years distant, we saw warm dust grains glowing as in rings of fire (figure 5.6).

For astronomers, these images provide insight into the exquisite structure of the universe and, I would argue, into God's artistry. Such is the never-ceasing excitement of being a researcher guided, step by step, by a Creator who wants to be discovered. God opened another research path in Australia when I worked as a guest of Professor Kenneth Freeman at the Australian National University (Canberra). Freeman and I have coauthored books not only on astronomical matters but on the nature of revelation and on the nature of truth. We have both been intensely interested in the famed two books discussed by Galileo: the book of nature (including observations using telescopes) and the book of

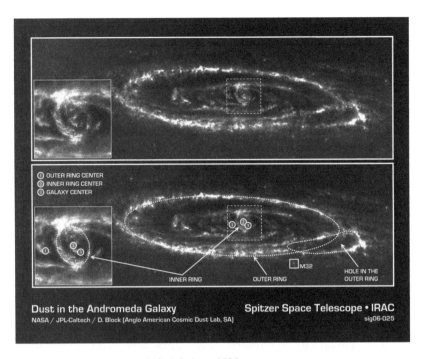

Figure 5.6: Andromeda Spiral Galaxy M31
Glowing warm dust grains in the Andromeda Spiral Galaxy M31, imaged using the
Spitzer Space Telescope (in orbit about the Sun), at a wavelength of eight microns.
Our images reveal "rings of fire"—the aftermath of a near head-on collision between
M31 and its neighbor, M32. *Credit: NASA/JPL/D. L. Block.*

Scripture. We affirm, with Galileo, that truth does not reside in one book alone.
The book of nature is a book of process, while the book of Scripture is a book
of purpose.[14]

To elucidate a little further: The *truth of nature* (medicine, astronomy, and
other academic disciplines) must be understood in light of the *nature of truth*,
which includes God's spiritual revelation to us as carefully unfolded in the book
of Scripture. We can know both with the help of God's Spirit.

Jesus likens the Spirit of God to a wind. In the Gospel according to John,
we read: "The wind blows wherever it pleases. You hear its sound, but you cannot tell where it comes from or where it is going. So it is with everyone born of
the Spirit" (John 3:8).

For some people, only what we can see or detect with our senses is the sum
of all reality. Winds of any trace of spiritual revelation are usually expunged

from discussion in my academic circles. Invoking a sense of purpose in the book of nature is, today, a revulsion to many. Such is the mood of our age. It seems that an "exclusion principle" seems to be at work (for example, embracing astronomy but no God behind it all). In a fallen state, humans simply cannot "see" the Spirit of God with our eyes,[15] yet He's the one who invites all of us to "see" what He's created—if only we will open our eyes.

Helpful Insight

Details of Galileo's infamous trial (1633) need not be repeated here, but in Pope John Paul II's 1992 apology to Galileo, he cited (translated from Latin) Cardinal Caesar Baronius (1538–1607): *"Spiritui Sancto mentem fuisse nos docere quomodo ad coelum eatur, non quomodo coelum gradiatur."*[16] In English, this translates, "It was the Holy Spirit's intent to teach us how one goes to heaven, not how the heavens go." How do the heavens go (or move)? How are stars born? How are black holes formed? Answers to such questions belong to scientific disciplines such as astrophysics and relativistic astrophysics. However, there we seek a description of *process*, not of *purpose*.

At the heart of the Gospels is the incarnation—God becoming flesh (figure 5.7), which is beyond human reason. Reason and revelation are not the same thing, but they are never opposed. Both expound truth. The book of Scripture cannot be in contradiction with the book of nature, provided our understanding of both books is correct—context is all important.

As the father of modern science, Galileo encouraged theologians to understand the contextual meaning of the Scriptures. For example, Jesus says "I am the door" (John 10:9). One dare not take such a statement as literally meaning that Jesus is an actual wooden door. Rather, His statement has an infinitely wider sweep of truth: it's the singular passage for His flock into His kingdom.

In that vein, when we consider a Bible verse such as "The sun rises and the sun sets, and hurries back to where it rises" (Ecclesiastes 1:5), are we to conclude that astronomy has revealed a geocentric universe, wherein all objects (including the Sun) orbit the earth, as argued by multitudes of theologians at the time of Galileo? Absolutely not. What kind of cosmos has modern astronomy then revealed? It shows a heliocentric universe where the earth and all other planets orbit the Sun. And our Sun is one star amongst trillions of stars.

The book of nature is encased in the language of mathematics—the foundations of cosmology invariably rest upon the Einstein field equations, for example, while the focus of the book of Scripture is a dynamic relationship between God and His people. Cardinal Baronius is correct. Theologians at

Figure 5.7: *Gloria in Profundis Deo*—**Glory to God in the Lowest**
The hinge between the two books—the world and the Word—is the incarnation. The house is bathed under the rays of the star of Bethlehem (Matthew 2:9–11). Alongside the house, a young person, enthralled by the star, aims a handheld telescope heavenward. At central focus in the house is the King of the Jews. *Credit: Gift of the Belgian artist Frans Claerhout to the author.*

the time of Galileo (or at any other epoch) cannot control the geography of the heavens (process). However, those heavens declare the glory of God to us (purpose).

References to the book of nature stretch back centuries. The following quote from St. Augustine (354–430) is explicit: "It is the divine page that you must *listen to*; it is the book of the universe that you must *observe*"[17] (emphasis added).

The book of Scripture is widely accessible in our (largely online) Western world, but it was not always so, even in English.[18] In contrast, extensive academic training has always been required to insightfully comprehend the book of nature. Even in that second book, however, progress is made over decades (and centuries) by scientific advance.

Figure 5.8: Solar Eclipse (May 1919)
Our closest star, the Sun, eclipsed by the Moon during the historic solar eclipse of
May 1919. The image unveils stunning details in the solar corona and a giant promi-
nence emerging from the upper right part of the Sun. Stars (circled) in the constel-
lation of Taurus (the Bull) were photographed in broad daylight and their positions
confirmed predictions of Einstein's general theory of relativity. Space-time in the solar
vicinity is curved. *Credit: ESO/Landessternwarte Heidelberg-Königstuhl/F. W. Dyson,
A. S. Eddington, & C. Davidson.*

Theological biases can't obliterate elegant truths in both Genesis 1 and in
cosmology, such as our expanding universe discovered by Belgian priest and
theoretical physicist Georges Lemaître in 1927.[19] He showed us that the universe had a beginning both in space and in time. What does this mean for
human beings?

Given the sheer immensity of our universe, some people might succumb
to a feeling of "being lost" in it. Are human beings not a mere speckle in the
cosmic fabric of space and of time? I would argue otherwise: there is grace in
space. I exist within a spatial framework of being positioned at a certain locale
at any specific time, but it was at a specific locale in *space* (Johannesburg, South
Africa) and at a specific *time* that I experienced God's grace flooding my soul,
over 45 years ago.

Finding the Light

I'll conclude by using the analogy of a total solar eclipse (figure 5.8). The brightest star in our sky—the Sun—may be eclipsed to such an extent that birds may start flying back to their nests due to the "encroaching night." And stars in the proximity of the Sun may be photographed in broad daylight. Why, then, this analogy of an eclipse?

We should never allow our minds to be stuck in the path of the darkness of a total solar eclipse—never. As Auschwitz survivor Dr. Edith Eger reminds us: "We can't choose to vanish the dark, but we can choose to kindle the light."[20] Blessing the Shabbat candles (*bentsch likht*)—to kindle their light—evokes in me the warmest of memories.

In this chapter, my thoughts in this essay as a Jewish astronomer have ranged from the universe of stars and of galaxies to the universe of the heart. In the words of French mathematician and physicist Blaise Pascal, "The heart has its reasons, which reason does not know."[21]

This is God's universe—one where grace prevails. In this universe, both reason and revelation deliver the same message. It should not surprise us that people who are trying to make sense of this world have been provided with a map by the Maker of this world, who, by His grace, has visited His world in person.

Chapter 6

Seeing the Designer in the Lab

Fazale "Fuz" Rana

Girls, rock music, and sports. Those were the three things that most captured my interest during my youth. And, unlike many of my classmates, I had no real idea what I wanted to do after high school, let alone what I should study in college.

Because I was a good student—I was the covaledictorian of my high school class—my father urged me to enroll in the premed program at West Virginia State College (WVSC). Even though I had very little interest in science and math as a high school student, I excelled in both. So, I thought, "Why not?" To my surprise, I received a full four-year Presidential Scholarship at WVSC.

My father wanted me to get a jump start on college, so I took an introductory course in biology during the summer before officially beginning school in the fall. Day after day for six weeks, I trudged up several flights of stairs to the top floor of the old science building to take Biology 101. The reward for reaching the top floor was sitting for hours on end in a hot, humid lecture hall and adjoining laboratory—without air-conditioning. The miserable stickiness I felt, however, was worth it. By taking this course, I unexpectedly stumbled upon the direction for my academic and professional future.

It all began with a simple, but profound, question: "What is life?" This question is usually one of the first topics addressed in an introductory course in biology. That makes sense. If someone is going to study life, then it's necessary to know exactly what is to be investigated. I was astonished to learn that scientists did *not* know how to define life. And we still don't. We can list the characteristics common to all life on Earth, but we can't define it. My surprise turned to curiosity. And that curiosity became an obsession. I hungered to know:

- What is life?
- What constitutes life and how does it operate at its most fundamental level?
- How did life begin?

Biochemistry is one of the disciplines that holds significant potential to answer these questions. I wanted to be a biochemist. I wanted to understand as much as I could about the fundamental features of life at its most basic level—the molecular level.

My interests would expand beyond girls, rock and roll, and sports to include chemistry and biology. The focus of my undergraduate studies shifted from preparing myself to go to medical school to doing everything in my power to get ready for graduate school and a PhD in biochemistry. Even as a college freshman, I began to dream about one day heading up my own lab and making important scientific discoveries.

Questioning God's Existence

As I took classes in biology, I learned the naturalistic evolutionary paradigm—the grand claim that all of biology, including the design, origin, history, diversity, and distribution of living organisms could be explained by natural evolutionary mechanisms. I embraced this idea. Why wouldn't I? It's the mainstream scientific explanation for origins in biology. At this point in my academic training, my convictions about the validity of the evolutionary paradigm weren't based on a thorough examination of the evidence but rather on the authority I granted to my biology professors. I admired my professors and uncritically accepted what they taught. They often implied that science and religion were at war with one another. They taught me to reject supernatural explanations for the world and I responded to their tutelage.

Embracing naturalistic evolution and rejecting supernatural causation felt comfortable to me. When I entered college, I was an agnostic. I didn't know if God existed and, honestly, I didn't care. Religion didn't interest me. That wasn't always the case, though. When I was in high school, I spent time exploring Islam. In fact, I recited the Shahada—the declaration that Allah is the one true God and Mohammad is his one true prophet. My father taught me how to pray to Allah and I read from English translations of the Quran.

My father was a devout Muslim. He prayed and read from the Quran every day and carried a prayer book with him in the breast pocket of his shirt. I became interested in Islam in my teenage years partly because I wanted to connect with my father and my heritage. However, after about a year or so, I gave up on Islam. The writings of the Quran felt esoteric and unclear to me. Prayer felt like an obligation and soon became a burden to me.

Abandoning Islam left me unsettled and unsure about God's existence. My naïve confidence in the evolutionary paradigm validated my agnosticism. I reasoned that if evolutionary mechanisms can account for the origin and history of life—and the design of biological systems—then what role is there for a Creator to play? I wasn't alone in that mindset. In his book *The Blind Watchmaker*, evolutionary biologist and atheist Richard Dawkins notes, "Although atheism might have been *logically* tenable before Darwin, Darwin made it possible to be an intellectually fulfilled atheist"[1] (emphasis original). Over the years I've interacted with many people who seem to be intellectually fulfilled atheists who regard the evolutionary paradigm as the best way to account for biology.

New Discoveries Changed My Mind

I was excited to start graduate school. Finally, I could immerse myself in the study of biochemical systems. I loved taking advanced courses in chemistry and biochemistry and was thrilled to begin my own research projects. I valued the time I spent discussing ideas in biochemistry with my professors and my fellow students. Through these activities, I soon became impressed with the characteristic elegance, sophistication, and ingenuity of biochemical systems. I began to wonder, "How did biochemical systems originate?" This question is at the heart of the origin-of-life problem that scientists seek to solve. Though not part of my formal coursework, I began to study the origin-of-life question whenever I had the chance.

I discovered that scientists who focused their research efforts on this question had explanations for life's start. Accordingly, complex chemical mixtures on the early Earth self-organized into the first living entities in a gradual, stepwise fashion—a process scientists call chemical evolution.

My initial investigation into chemical evolution left me unsatisfied. The mechanisms of chemical evolution seemed to be entirely inadequate to produce the cell's vastly complex, highly sophisticated, tightly orchestrated chemical systems. Based on my experience as a chemist, I knew that chemical systems could self-organize, but the organization displayed by biochemical systems differed qualitatively from the order possessed by crystals and other types of molecular aggregates that form spontaneously.

Even the best chemists struggle to get a few chemicals in a flask to do what they want, even when they expend enormous mental effort and rely on the past work of other chemists. How could random physical and chemical events yield the amazingly elegant and highly integrated biochemical systems found inside the cell?

One of the primary goals of a graduate education is to teach students to think independently through scientific evidence and to develop conclusions based on the evidence alone, regardless of what other experts might say. Because I was learning to think for myself, I was willing to ask questions that I had not voiced as an undergraduate. I was also willing to follow the evidence where it led. The elegance and sophistication of biochemical systems combined with the significant scientific challenges confronting chemical evolution led me to conclude that life's beginning required more than chemical and physical processes alone. I reasoned that life's origin must have a supernatural basis. It must be the work of a Creator. And for the first time, I recognized that a rapprochement exists between science and religion.

Learning the Creator's Identity

Convinced that a Creator must exist, I began to privately ponder the Creator's identity. Initially I thought all the different religions led to the same truth—that God revealed himself to humanity in different ways, at different times. Religious pluralism made sense to me because when it came to morality, I saw overlapping truth claims among the world's great religions. Naïvely, I failed to recognize the substantial disagreements that also exist among the different religions about the nature of reality and God's qualities.

As I pondered questions about who God is and how I relate to him, if at all, my fiancée, Amy, rededicated her life to Christianity. She began sharing her renewed faith with me. I reacted with indifference. I accepted her decision to pursue Christianity, but I didn't think it was for me. I felt more comfortable embracing some type of universalism.

A few months later, the pastor who eventually officiated our wedding challenged me to read the Bible by appealing to my pride as a scientist. He reminded me that a scientist should be willing to seek after truth no matter where the search led him. I felt the thrust of his point and decided for Amy's sake that I should at least examine Christianity. This was the first time in my life that I seriously read and contemplated the Bible. The Gospel of Matthew convinced me that Jesus was who Christians claimed him to be. The Sermon on the Mount forced me to recognize my sins and need for a Savior. Jesus was the Creator I discovered inside the cell!

The elegance of life's chemistry led me to recognize the Creator's existence.

The religious experience I had while reading the Bible inspired my conversion. I embraced the Christian faith not because of my need for a "crutch." I came to faith in Christ as a scientist pursuing an understanding of biochemistry.

Can Science Detect a Creator's Handiwork?

I am a scientist. And as a scientist I recognize that all conclusions in science are provisional. They can change as we make new discoveries. Since my graduate school days, I have continually monitored the latest evidence from biochemistry and origin-of-life studies. I am more convinced than ever that the elegant design of biochemical systems can only be explained as the Creator's handiwork.

Still, one thing troubled me. My conclusion that a Creator played a role in the origin of life was based on intuitive cognition. But that type of conclusion isn't enough for a scientist. Scientific convictions must be based on *reflective* cognition.

In that vein, over the last three decades I have pursued the goal of turning my initial intuition of design into a robust scientific case for a Creator. On the surface, a case for design may seem impossible because it violates the tenets of methodological naturalism (the philosophical position that scientific explanations must be restricted to natural processes). The way forward came to me in the fall of 2017 as I was preparing for a debate with Dr. Stanley Cohen, a biologist from DePaul University. Our debate, held at Olive–Harvey College in Chicago, centered around the question, "Is there evidence for a supernatural creator in nature?" I realized that at least a few scientific disciplines are predicated on the capacity of science to detect the work of intelligent agency in nature and distinguish that activity from the outworking of natural processes. This insight means that it is possible to make a case for a Creator, while conforming to the constraints of methodological naturalism.

As one example, forensic scientists can determine if an individual died as the result of natural processes, by accident, or by the intentional action of another person—a malevolent intelligent agent. Second, in the quest to identify alien civilizations, researchers at the SETI (Search for Extraterrestrial Intelligence) Institute monitor electromagnetic radiation emanating from distant astronomical objects to look for signatures that bear the hallmark of intelligent agency. In the early 1970s, chemist Leslie Orgel and molecular biologist Francis Crick proposed directed panspermia to explain the origin of life on Earth, and they even suggested ways to scientifically test this idea.[2] This idea asserts that an alien civilization intentionally seeded Earth with life. In a third example, anthropologists examine pieces of rock to determine whether the stones were intentionally fabricated into a tool by a hominin (such as Neanderthals) or merely shaped by natural processes.

Indeed, science does possess the tool kit to detect the work of an intelligent agent and distinguish it from natural causes and events. If so, then why can't scientific inquiry determine if an intelligent agent played a role in the origin, history, and design of life and the universe?

A Methodology to Detect the Work of Agency

As a scientist I know that before proceeding with any scientific investigation, valid methods must be in place. What scientific methods can we use to determine whether a Creator played a role in the origin and design of life? As I pondered this question as part of my debate preparation, I realized I could study the methods anthropologists use when they assess whether an artifact or the features of an archaeological site show the deliberate activity of a hominin or not. This same methodology can be used to gain insight into the genesis of biochemical systems.

Fortuitously, it was at that time that the Cerutti Mastodon site outside of San Diego, California, made the news. This site was unearthed during the early 1990s when road construction crews unintentionally uncovered the remains of a single mastodon. Though the site was excavated from 1992 to 1993, scientists were unable to age-date the remains. Both radiocarbon and luminescence dating techniques failed.

Then, in 2017, researchers turned failure into success by using uranium-series disequilibrium methods to date the site at about 130,000 years old.[3] This result shocked the team because analysis at the site indicated that the mastodon remains were deliberately processed by humans. But modern humans hadn't made their way into the Americas until around 15,000 to 20,000 years ago. This means that other hominins must have arrived in the Americas before modern humans, with the most likely candidate being Neanderthals.

Some anthropologists remain skeptical that Neanderthals—or any other hominin—modified the mastodon remains.[4] Why? Because this claim defies the conventional model for the populating of the Americas by humans. This claim also fails to address the observation that the sophistication of the tool kit used at the Cerutti site doesn't match that produced by Neanderthals 130,000 years ago based on archaeological sites in Europe and Asia.

So, did Neanderthals make their way to the Americas 100,000 years before modern humans? An interesting debate has ensued and will most certainly continue in the years to come. My point isn't to adjudicate this debate. In fact, it doesn't really matter to me if this claim stands. I realized that the methodology archaeologists employ to detect the work of agency in shaping features found in nature is beautifully illustrated by the anthropologists who sought to justify their claim that Neanderthals were present in North America 130,000 years ago. Because of the controversial nature of their claim, the research team presented their case carefully, which consisted of three interrelated categories of evidence.

The Appearance of Design. The archaeologists argued that: (1) the unnatural arrangement of the bones and stones and (2) the markings on the stones and the fracture patterns on the bones appear to result from the intentional activity of a hominin. Specifically, the mastodon bones displayed spiral fracture patterns that looked as if a creature, such as a Neanderthal, struck the bone with a rock—most likely to extract nutrient-rich marrow from the bones. To put it another way, the archaeological site shows the appearance of design.

Failure to Explain the Evidence Through Natural Processes. The archaeologists explored and rejected alternative explanations for the arrangement, fracture patterns, and markings of the bones and stones, such as scavenging by wild animals. In a follow-up study, anthropologist Ruth Gruhn even ruled out that the breakage of the mastodon bones was caused by heavy machinery used at the site before Caltrans unearthed the mastodon remains.[5]

Reproduction of the Design Patterns. The archaeologists confirmed—by striking elephant and cow bones with a rock—that the markings on the cobble and the fracture patterns on the bone were made by a hominin. They took elephant and cow bones and broke them open with a hammerstone. In doing so, the researchers produced the same type of spiral fracture patterns in the bones and the same type of markings on the hammerstone as those found at the archaeological site. That is, through experimental work in the laboratory, they demonstrated that the design features were, indeed, produced by intelligent agency.

Making a Scientific Case for a Creator

Using the same criteria as anthropologists, I've unwittingly made a scientific case in three of my books that life's origin and the structure and function of biochemical systems are best explained as the outworking of intelligent agency.

In *The Cell's Design*, I argue that the analogies between biochemical systems and human designs evince the work of a Mind and serve to revitalize philosopher William Paley's Watchmaker argument for God's existence. In other words, biochemical systems display the appearance of design.

In *Origins of Life*, Hugh Ross (my coauthor) and I demonstrate the deficiency of natural process, mechanistic explanations (such as replicator-first, metabolism-first, and membrane-first scenarios) to account for the origin of life and, hence, biological systems. Currently, there is no compelling natural process explanation for the origin of life and, hence, the origin of biochemical systems.

In *Creating Life in the Lab*, I describe how work in synthetic biology and prebiotic chemistry empirically demonstrates the necessary role intelligent agency plays in transforming chemicals into living cells. When scientists go into the lab and create protocells, they are demonstrating that the design of biochemical systems can only come about through the work of a Mind.

Space limitations prevent me from elaborating on each of these criteria. Suffice it to say, contrary to widespread opinion, we can reasonably conclude—based on an objective scientific analysis and resembling the approach used by anthropologists—that a Creator played an integral role in the origin and design of life's most basic, fundamental systems. My intuition about the Creator's role in the origin and design of life's most basic systems became a *bona fide* scientific conclusion.

But my journey didn't come to an end after my debate at Olive–Harvey College. It continues to this day. More recently, I have come to recognize that the case for a Creator can be further affirmed by simultaneous advances in biochemistry and nanotechnology. The latest insights from these disciplines highlight the arresting similarity between the technology we invent as human designers and the architecture and operation of biochemical systems. Biochemistry is nanotechnology.

The Relationship Between Biochemistry and Technology

As I discussed, central to the scientific case for a Creator is the appearance of design displayed by biochemical systems. Over the last several decades, biochemists have gained understanding about the structure and function of these systems. Provocatively, this work takes us beyond nonspecific qualities such as sophistication and elegance to reveal an eerie, precise similarity between the designs of biochemical systems and those characteristic of human technology. To put it another way, when humans design, invent, and create systems, objects, and devices, the things we produce have certain telltale features that reflect agency. It is remarkable to see that the hallmark properties of biochemical systems display these same telltale design signatures. For reference, I delineate these similarities in my book *The Cell's Design*.[6]

These shared features have led to an interesting and productive interplay between biochemical systems and technology. This relationship helps biochemists develop an understanding of the architecture and operation of biomolecular systems by comparing them with human technology. Conversely, engineers take advantage of this relationship by using the structure and function of biochemical systems to inspire technological innovations.

This interplay also allows us to construct three versions of the Watchmaker argument for God's existence and role in the origin and design of life. I have dubbed these variants (1) the revitalized Watchmaker argument, (2) the converse Watchmaker argument, and (3) the Watchmaker prediction.

The Watchmaker Argument

I have come to admire William Paley, an eighteenth-century Anglican natural theologian. In my opinion, his genius is often overlooked. Paley proposed his well-known Watchmaker argument in 1802. He posited that just as a watch requires a watchmaker, so too, life requires a Creator. In the opening pages of *Natural Theology*, Paley sets the stage for his famous analogy by comparing a stone to a watch:

> But suppose I had found a watch upon the ground, and it should be inquired how the watch happened to be in that place; I should hardly think of the answer which I had before given [for the stone], that for any thing I knew, the watch might have always been there. . . . When we come to inspect the watch, we perceive (what we could not discover in the stone) that its several parts are framed and put together for a purpose . . . the inference, we think, is inevitable, that the watch must have had a maker: that there must have existed, at some time, and at some place or other, an artificer or artificers who formed it for the purpose which we find it actually to answer: who comprehended its construction, and designed its use.[7]

In Paley's day, a well-made watch exemplified expert craftsmanship. Paley pointed out that a watch is a contrivance—a machine composed of several parts that interact precisely to accomplish its purpose. He then contrasted a watch with a rock. A rock, Paley argued, finds explanation through the outworking of natural processes; but a watch requires a mind to explain its existence.

Based on an extensive survey of biological systems, Paley concluded that the properties of living systems have more in common with a watch than with a rock. And if a watch requires a human watchmaker to account for its existence, then living systems require a Divine Mind to explain their origin.

The recently recognized similarity between biochemical systems and human technology allows us to present a reinvigorated version of Paley's argument. The revitalized Watchmaker argument is nicely illustrated by the discovery of proteins that operate as biochemical Turing machines when they interact with DNA.

The Revitalized Watchmaker Argument

In recent years computer scientists and molecular biologists have come to realize that the cell's machinery, which manipulates DNA, literally functions like a computer system at its most basic level of operation. I find it mind-boggling. To fully appreciate this insight, we need to consider the work of British mathematician Alan Turing.

Turing is considered by many to be the father of modern-day computer science due to the groundbreaking theoretical work he performed prior to World War II. He conceived of abstract machines, known today as Turing machines, that consisted of an input, finite control, and an output.

Remarkably, the cell's machinery manipulates genetic information in precisely the same way that Turing envisioned his imaginary machines to operate. Information housed in DNA is digital information. Whenever a complex biochemical process (such as DNA replication, transcription, or DNA repair) takes place, the cell's machinery is essentially carrying out a computer operation.

It should be obvious how this insight bolsters the Watchmaker argument. The incredible power of computers ranks as the pinnacle of human engineering achievements today. And we know that computers have been designed and assembled by intelligent agents. So, why would we conclude anything else about the biochemical computer operations centered around DNA?

The Converse Watchmaker Argument

One of the most exciting areas of engineering and technology is the field of biomimetics and bioinspiration. Here, researchers deploy biological and biochemical designs to inspire new technologies and to solve engineering problems. DNA computing serves as a case in point.

The similarity between cellular processes—such as transcription, DNA replication, and DNA repair—and the fundamental operation of computer systems has given inspiration to an exciting new area of nanotechnology called DNA computing. The brainchild of Leonard Adleman, DNA computers consist of DNA and the proteins that manipulate this biomolecule inside the cell. These computers are housed in tiny test tubes, yet they are more powerful than the most powerful supercomputer system humans have devised. That power stems largely from the capacity to perform a vast number of parallel computations simultaneously.

Researchers have used DNA computers to solve problems that silicon-based supercomputer systems cannot solve because of their limited computational power compared to DNA computers. Bioinspired designs such as DNA computers allow us to formulate a variant of the Watchmaker argument that I dub the *converse* Watchmaker argument. Accordingly, if biological designs are the work of a Creator, then these systems should be so well designed that they can serve as engineering models and inspire the development of new technologies. In this way, the disciplines of biomimetics and bioinspiration add support for the Watchmaker argument. However, we wouldn't expect the same to be true of these systems if they were generated by an unguided, historically contingent natural evolutionary process that modified existing designs and cobbled them together to create new designs.

The Watchmaker Prediction

In conjunction with my presentation of the revitalized Watchmaker argument in *The Cell's Design*, I also proposed the Watchmaker *prediction*. I contend that many of the cell's molecular systems currently go unrecognized as analogs to human designs because the corresponding technology has not yet been developed. That is, the Watchmaker argument may well become stronger in the future and its conclusion more robust as human technology advances. This prospect leads to the Watchmaker prediction: As human designers develop new technologies, examples of these technologies, which previously went unrecognized, will become evident in the operation of the cell's molecular systems. In other words, if the Watchmaker argument truly serves as evidence for the Creator's existence, then it is reasonable to expect that life's biochemical machinery anticipates human technological advances.

Fulfillment of the Watchmaker prediction is illustrated by advances in nanotechnology such as those centered around Brownian ratchets. One of the chief technical hurdles that stands in the path to viable nanodevices is the inability to generate directional movement within nanomachinery. Some researchers have proposed Brownian ratchets as a way around this technical barrier.[8]

Brownian Ratchets

Brownian ratchets refer to devices that take advantage of Brownian motion. This phenomenon describes the random, zigzag movement of microscopic objects suspended in a liquid or gas. In 1827, botanist Robert Brown noted that pollen particles dispersed in water dart about in a zigzag fashion, now referred to as Brownian motion. Brownian motion stems from the net force exerted on a suspended object by the gas or liquid molecules. Occasionally, the sum of these forces generates a directional force that causes the particle to move. This force is short-lived. When the directionality of the force ceases, the particle stops moving. Overall, the influence of Brownian motion causes the particle to move in a random, jerking, zigzag fashion.

Brownian ratchets exploit Brownian motion but use barriers to restrict the motion, so it only occurs in a specified direction. Brownian ratchets require energy input to erect and maintain the barriers that prevent motion in unwanted directions. Technologists have hoped that Brownian ratchets could serve as the means to generate directional movement in nanodevices. This dream became reality in 1999, when a research team built a device that transports DNA molecules using a Brownian ratchet.[9]

While nanotechnologists continue to look for ways to develop and implement Brownian ratchet technology in nanodevices, biochemists have recently discovered several Brownian ratchets inside the cell. For example, some of the biochemical Turing machines that operate on DNA make use of a Brownian ratchet to power their movement along DNA. One of these proteins is T7 RNA polymerase.[10]

After the discovery that T7 RNA polymerase operates as a Brownian ratchet, biochemists have come to recognize that DNA polymerases (which play a central role in DNA replication and repair) also operate as Brownian ratchets as they move along DNA.[11] Without the invention of Brownian ratchets, proteins like T7 RNA polymerase and the DNA polymerases wouldn't be included in the collection of biochemical systems that contribute to the revitalized Watchmaker argument. I find it provocative to learn that as researchers develop new technologies, examples of these technologies *already exist* in the biochemical systems that constitute life. It's no wonder that the structure and function of biochemical systems serve as such a suitable muse for engineers and technologists.

Are Science and Faith in Conflict?

One of the most common narratives in our culture today is that science and faith are at war with each other. Yet, except for a brief period during my college years, I have never thought that to be the case. Scientific insights I learned about biochemistry and the origin-of-life question as a graduate student convinced me that the rumors of peace are true, and a Creator must exist.

As a Christian today, I don't find my convictions to be out of the ordinary. Both the Old and New Testaments teach that God has revealed himself to us through creation—the record of nature. If so, then we should expect that scientific inquiry will uncover evidence of God's fingerprints. And, indeed, the Creator's signature is evident in the structure and function of biochemical systems.

This conclusion is more than a theological or a philosophical conclusion; it is rooted in science. Using the same criteria that an anthropologist would use to study the archaeological record, I conclude that life's origin and the fundamental design of living systems are best explained by the work of an Intelligent Agent. This insight has practical implications that provide a framework for understanding biochemical systems and grounding work in biomimetics and bioinspiration. If a Creator is responsible for the design of biochemical systems, then these systems should point the way for new technologies. Indeed, they have and will continue to do so.

Chapter 7

A Vision for Archaeology and Faith

John A. Bloom

Had I been born a few centuries ago, I would have been a beggar or a jewel-er—most likely the former. I was very nearsighted, which oddly wasn't di-agnosed until I was six years old, but eyeglasses would dramatically change my view of the world. Suddenly things more than a couple of feet away were clear and sharp! Rather than asking for alms or crafting intricate jewelry, I could now lead a fairly normal life, thanks to my Coke-bottle lenses.

This early limitation, coupled with my personality, fueled an intense in-terest to see things up close and to figure out how they worked. I remember my father carefully breaking the glass off a 3-way light bulb to show my older brother and me the two filaments inside. The bulb's 50/100/150-Watt bright-ness depended on whether one or the other or both filaments were lit. It was incredible!

I didn't see well enough to play much with others—aside from my broth-er—so I spent most of my young days building things with Tinker Toys, an Erector Set, and plastic bricks. I also played with earth-moving toys in our garden.

Fortunately, and probably to keep me out of trouble, my third-grade teach-er Miss Bramhall gave me an astronomy project. She asked me to paint the ob-jects in the solar system and she posted them above the classroom chalkboard. Although astronomy pictures at that time were mostly fuzzy black-and-white images of planets and galaxies, the project opened the universe to me and ce-mented my interest in the sciences. A chemistry set for a Christmas present and my overall passion for anything electrical or electronic also definitely helped.

Uneasiness in College

In college, I followed in my brother's footsteps and declared a physics major, although I also loved chemistry and added that major in my sophomore year. Thankfully, I was good enough at math to manage physics, although I preferred the hands-on experiments over the calculation-heavy theoretical area. Thanks to my extreme nearsightedness, I had no trouble wiring tiny circuit boards and building equipment with small parts.

The complete shift in academic and social worlds that accompanied college life also affected me spiritually. Growing up, my family had attended a small, conservative Presbyterian church and we had a great pastor. I memorized Bible verses and parts of the Westminster Confession of Faith in catechism class and had been confirmed by the congregation. Christianity seemed reasonable. It had an intellectual foundation and I didn't have any problems "believing" it, but the Christian life just seemed like it was a series of mental boxes to check off. So, in college, religion in general and Christianity in particular just didn't seem to matter. Thankfully, God kept me from doing anything crazy (no frat life) and I focused mainly on academics.

Coming into my junior year, I experienced some strained relationships, including one where my girlfriend dumped me for my roommate. After that punch to the gut, I took the risk of asking the prettiest girl in my biochemistry class if she would be my lab partner. She took the risk of accepting me, and that fall semester we struggled together through a series of hard labs that usually didn't work (this was the first time the course and lab were being taught, and we were the "shakedown" class). We worked well together, but no, we didn't get married later. However, I noticed a difference in her. She was *really* a Christian: patient, graceful under pressure, not anxious or stressed out over failure like everyone else was, including me. I had to ask her: "What do you have that I don't?" Her reply: "You have a Bible, don't you? Go and read it."

So, I did. It wasn't strange material, but now I was seeing something I had missed before. We all seek peace and security, but I was looking for these in all the wrong places: in relationships with other broken people and in material things that wear out or get stolen. Basically, I'd missed the security of a relationship with my Creator by instead focusing only on the things he'd created. Like the woman who'd spent twelve years and all that she had on physicians in the hope of getting healed (Mark 5:25–34), I realized that I needed to reach out and have a relationship with Jesus, not just give mental assent to some details about his life and character.

Challenged by My Peers

But when I started to talk with my friends about a relationship with Jesus, they threw tons of questions at me, like: Hasn't science shown that God doesn't exist? What evidence is there that anything in the Bible could be true?

Growing up I had seen Christianity and science as partners. It seemed clear that the universe had a beginning, that the Genesis narrative had the basic sequence of creation right, and that Scripture described our flawed human condition (I saw Adam and Eve fitting into a God-guided theistic evolutionary framework at the time). I didn't see science as a barrier to faith at all. But my friends' questions were reasonable: Could I trust the Bible itself? What about the details regarding Israel's history and the New Testament accounts of Jesus and the apostles? All their questions boiled down to three points: Did it happen? Was it true? Does it matter?

By this time, I was finishing college and was privileged to be accepted into the graduate physics program at Cornell University. There, through the "chance" recommendation of a college friend, I connected with a student-centered church that focused on apologetics. This church provided resources to answer not only science-faith questions but also the historical and biblical reliability challenges. In addition to his regular teaching, the pastor recommended a list of "must read" books to address these issues.

I quickly devoured Frank Morison's classic, *Who Moved the Stone?* Morison set out to disprove the resurrection, but after careful examination of the historical sources, he came to accept its historicity.[1] Similarly, William Ramsay's *St. Paul the Traveller and the Roman Citizen* reflects the careful work of a skeptical historian who came to accept Acts and Luke as accurate accounts of Jesus's ministry and the early church.[2] (Both books are still in print with updated editions, although Josh and Sean McDowell's *More than a Carpenter* covers highlights of this material and is more accessible to contemporary, casual readers.)[3]

But Morison and Ramsay appeared to be the exceptions. Most Bible scholars in the academy today remain skeptics. Thus, when I finished my physics degree, I thought it best to go deeper into biblical history myself by attending seminary and then pursuing graduate studies in the Old Testament (OT) because historical questions, in addition to science-related questions, focus much more on OT texts (New Testament issues are comparatively minor).

What My Studies Revealed

When I got into graduate OT work, I encountered a somewhat different approach than what I had found in physics. New data was scarce. You couldn't

easily set up a lab experiment to test something about the text. But more striking to me than the data limitation was the common attitude that almost every
statement in the Bible was assumed to be wrong! And what data could counter
this skepticism? Only material from outside the Bible, mainly from archaeology. Thus, if I wanted to work on the historical reliability of the Bible, I needed
affirmation from extrabiblical sources. So, I changed programs over to ancient
Near Eastern (ANE) studies in order to focus on the extrabiblical data that
most OT scholars and others judged to be more trustworthy.

But I learned something important about science along the way. At
Cornell, it had been drilled into me that as scientists, we were to seek truth,
no holds barred. What my thesis advisor or others expected the results to be
didn't matter. The only thing that counted was a well-controlled, unbiased as-
sessment of the data. But in a Hebrew language class when we were studying
the Genesis flood narrative, I was told that "in order to be scientific," we could
not accept the account as historical at all. It's strange to think that I earned a
PhD in physics from an Ivy League University and had never been told that
"science can only give naturalistic answers." This Bible class was my first exposure to this thinking. Had I slept through the Cornell class when this maxim
was explained? Perhaps, but scientists rarely study philosophy, and when one's
research focuses on studying material causes for material effects, it's a quest
for truth only. Had I been working in an origins-related science area where a
materialistic (or naturalistic) explanation was presumed before one even considered the data, my experience might have been different. Nevertheless, I was
shocked to learn that "science" meant the pursuit of materialistic explanations,
not the quest for truth.[4] And since many OT scholars want to be "scientific,"
their goal also is to explain away any miracles or special historical events in the
Bible as legends or misunderstood natural phenomena. Affirming that they actually might have happened, as Morison and Ramsay had come to believe and
argued, was simply unprofessional.

The ancient Near Eastern courses were exciting. They consisted of real ar-
tifacts, a range of literary, religious, economic, and administrative texts to de-
cipher and understand, and occasionally new discoveries to wrestle with. ANE
studies filled in the cracks for me and provided a cultural setting for the world
of the Bible (although the field has grown so large that much of the scholarly
work in Assyriology and Egyptology today has no direct bearing on biblical
material). Again, I saw the same efforts to "be scientific" and explain away anything miraculous, but I also noticed more openness to accepting at least a historical kernel behind biblical events.

So, did OT and ANE studies help me answer my three big questions regarding Bible history: Did it happen? Was it true? Does it matter? Most definitely! Those studies provided both evidence and the framework to interpret it. Let's look at three components of the framework first.

Materialism Is Pervasive

Finding truth requires digging through more than data. We all have philosophical assumptions that guide our thinking and incline us to favor some conclusions over others. In our world today, materialism (or naturalism) is the dominant perspective. It's the water in which we, like fish, swim. Thus, for many people this environment determines the outcome, and the data doesn't matter. Realizing the power of this embedded cultural worldview—both inside and outside of the hard sciences—allowed me to see that two people can look at the same facts and come to different conclusions. One can even claim to be following the evidence where it leads but remain unwilling to take the correct path because one has overriding assumptions against it. Thus, any discussion of historical and scientific data must include a conversation over what is an "acceptable" interpretation, and why.

Evidence Is Fragmentary

Scientists can never gather all the data and evidence that we want. In astronomy, for example, we can only sample the electromagnetic (and now gravitational) waves and occasional high-energy particles that course through the universe and that happen to impinge on the tiny earth. We cannot travel to other stars and galaxies to collect physical samples or climb into a time machine to go back and observe the big bang.

This limitation also applies to ancient history and archaeology. One can only hope to recover a fraction of the original material culture that survived destruction through burning and looting *and* did not decay as it was buried for thousands of years. And of these meager remnants, only a fraction will just happen to be discovered, properly analyzed, and published. This constraint gives the skeptic a huge advantage in biblical questions because we can only hope to recover a tiny bit of the ancient materials that relate to the text. However, since the skeptic has assumed that virtually none of the Bible is true, we can look at the *trend* in the data: Do new discoveries, as rare as they are, tend to confirm the biblical accounts, or undermine them? Thus, archaeology is not able to *prove* the Bible, but it can give us increasing assurance that what is reported there actually happened.

Belief Thresholds

Some Christians seem to think that the more faith they have, the better. I've known skeptics who have said that they wouldn't believe in Jesus unless he appeared to them physically, like he did to Thomas. What is the balance between proof and faith? Jesus did miracles to confirm his spiritual claims, such as heal the paralyzed man (Mark 2:1–12), so he's certainly not asking us to have blind faith in him. But how much evidence that is convincing to us personally should we expect God to provide? I think everyone has different thresholds, but my sense is that God provides everyone with enough evidence that they *can* believe in him (cf. Romans 1:20), but not with so much evidence that they *have to* believe in him. Faith is thus a choice to trust God, based on reasonable, but not overwhelming, evidence that he exists and is trustworthy. God apparently does not want coerced love, faith, and trust.

These three points gave me a framework to understand the data and its significance. It also gave me some sympathy for modern skeptics. In the remainder of this chapter, I'll present evidence from archaeology that gives us reason to trust the Bible, the God behind it, and the salvation that he offers.

The Resurrection

The physical, bodily resurrection of Jesus on Easter morning is the heart of the Christian message. Thus, it is striking that most liberal and critical scholars today grant that the apostles and the early church *believed and taught* that Jesus had physically risen from the dead, although they deny that it actually happened (remember, in order to be "scientific," scholars today must explain away miracles). This position is considerably more conservative than that of liberal skeptics decades ago who held that Jesus's resurrection and divine nature were myths that grew up around the "Jesus cult" some centuries later. The primary reasons for this shift are the early dates now accepted for the Gospels and book of Acts (following the work of Ramsay and others), and the recognition of simple creedal statements (easy-to-remember phrases expressing core beliefs) embedded in the Gospels and Epistles, that appear to predate the New Testament itself. For example, Paul's first letter to the Corinthians states,

> For what I received I passed on to you as of first importance: that Christ died for our sins according to the Scriptures, that he was buried, that he was raised on the third day according to the Scriptures, and that he appeared to Cephas, and then to the Twelve. After that, he appeared to more than five hundred

of the brothers and sisters at the same time, most of whom are still living, though some have fallen asleep. Then he appeared to James, then to all the apostles (1 Corinthians 15:3–7).

Gary Habermas is a modern-day Frank Morison who has devoted his career to assessing the evidence for the resurrection. His discussion and resources on early creeds are readily available.[5] Habermas is also well known for his "minimal facts argument," which builds a strong case for the historicity of the resurrection based on ancient secular sources and biblical sources that almost all liberal scholars today will accept as genuine (for example, that Paul actually wrote Romans and 1 Corinthians).[6]

The Empty Tomb

Is there archaeological evidence for Jesus's death and resurrection? During recent renovations at the Church of the Holy Sepulchre in Jerusalem, researchers exposed what appears to have been the actual stone slab where Jesus's body was laid,[7] which adds weight to the authenticity of the site.

Crucifixion

Was Jesus crucified? Although Roman crucifixion is well attested and described in ancient written sources, only three skeletal examples survive to illustrate the brutal practice. Although the remains are difficult to interpret, they show at least that large nails were driven through the ankle bones of the victims to anchor them to wooden beams.[8]

Early Christian Burials

Is there archaeological evidence for a Christian community living in Jerusalem, as described in the book of Acts? Jerusalem was densely populated in Herodian times and until its destruction in AD 70. To save space, it became customary during this period to transfer the bones of the deceased to ossuaries ("bone boxes") after their bodies had decayed away. Usually quarried from local limestone, ossuaries were often decorated and commonly had the name of the person(s) interred, their family relationship(s), and other notes chiseled on the box. Interestingly, ossuaries with crosses and other Christian symbols were found at three sites near Jerusalem: The Mount of Offence near Bethany, the tomb of Talpiot south of Jerusalem, and the Dominus Flevit site on the Mount of Olives.[9] The Talpiot ossuaries are clearly datable to before AD 70, they bear Jewish names, and in addition to some bearing the cross symbol, two boxes appear to have laments saying, "Jesus, woe!" on them. E. L. Sukenik, the famous Jewish archaeologist who excavated the Talpiot tombs, concludes his report on these finds by noting that the tomb was apparently in use until the middle of the first century AD (certainly not later than AD 70). He said, "All our evidence indicates that we have in this tomb the earliest records of Christianity in existence. It may also have a bearing on the historicity of Jesus and the crucifixion."[10]

These burial evidences of a Jewish Christian community in Jerusalem within decades after Jesus's death, plus the now generally accepted early dates for the Gospels, and the embedded creeds in the New Testament (NT) strengthen the case that "something revolutionary" happened in a tomb outside Jerusalem early one morning about 2,000 years ago.

The Great Isaiah Scroll

Reports and followers' beliefs about someone rising from the dead may be interesting, but what is the significance and meaning of such an event, if it actually did happen? This is where the greatest archaeological discovery of the twentieth century, the Dead Sea Scrolls, comes into play.[11] Among the thousands of manuscripts and manuscript fragments recovered in Qumran beginning in 1947, the most striking is the Great Isaiah Scroll, retrieved virtually intact, which contains the complete book of Isaiah and which carbon dates back to at least 100 BC.[12] What stunned scholars was that the text of the Isaiah scroll is essentially identical to the next-earliest Hebrew copy of Isaiah known, a codex dating to about AD 1000. This accurate preservation of the text for over a millennium confirmed Jewish scribal traditions about the extreme care they took in hand copying OT documents.

Here's the significance. Chapter 53 in the book of Isaiah presents the most detailed and extended passage about the torture, death, and exaltation of a future "suffering servant" who would die for the sins of the Jewish people. This prediction, dating back to the time of Isaiah and Hezekiah, who lived around 700 BC, set the Jewish cultural expectation of a messiah/redeemer who would someday come to rescue his people through substitutionary atonement. When John the Baptist pointed at Jesus and said, "Look, the Lamb of God, who takes away the sin of the world!" (John 1:29), everyone knew what he was talking about. Jesus did not arrive and act in a historical vacuum. A culture was prepared centuries in advance to expect him. Thus, the resurrection fits into a broader historical narrative where we are told its meaning and which we can document to well before Jesus's lifetime. The Great Isaiah Scroll is proof that this prediction was not edited or massaged by either Christians or Jews at a later date.

Seals of Hezekiah and Isaiah

In 2018, Eilat Mazar reported the discovery of clay bullae (round seals) bearing impressions from the seals of King Hezekiah and the Prophet Isaiah, a few feet apart from each other in a royal fortress/palace area of ancient Jerusalem.[13] These bullae are signature seals, a common form of identification and security at the time, where a small lump of wet clay was applied to a jar cover or to the string tying a document, and then stamped with one's personal seal. During a destructive fire the clay is heated and becomes a ceramic, and thus is preserved, while the papyrus or skin document or jar cover is burned. To have direct evidence of a seal that one can hold in one's hand from a famous biblical character who lived 2,700 years ago has the powerful impact of making ancient history come alive!

"Throwaway Details"

Archaeological evidence exists for many other kings of Israel and Judah, such as Jehu, Ahaz, and Jehoiachin, but a skeptic would say that it's not surprising to recover things relating to the leaders of a country. However, if the stories about what they did are legends, then one would expect the side characters in these tales generally to be fictitious. Thus, one recent discovery is striking. In 2007, Professor Michael Jursa discovered a Babylonian tablet in the British Museum archives. It was a temple gift receipt from Nebo-Sarsekim, the chief eunuch of Nebuchadnezzar II, dated to 595 BC. This is the same official described in Jeremiah 39 who was present at the fall of Jerusalem in 587 BC. As one scholar notes, that the Bible gets these "throwaway details" right when telling the story adds considerable weight to the truthfulness of the narrative.[14]

Seeing the Evidence

In archaeology we can only hope to recover a fraction of a fraction of a fraction of the original material culture from biblical times. While the skeptic might take comfort in the fact that very little of the Bible can thus be confirmed directly through archaeology, the trend in what has been recovered should give the skeptic serious pause. If the tiny fraction that we are finding supports the biblical account—implying that the biblical authors got the "throwaway details" right—then certainly this suggests that they got the main points right as well. Thus, I find that archaeology gives me sufficient evidence that I *can* believe the Bible, but not so much evidence that it *forces* someone to believe it. As mentioned earlier, this appears to be in keeping with God's character, as genuine love cannot be forced.

But I also find something else. I prefer to live with "the least amount of faith," if you will. No one can *prove* that their position is correct. The skeptic needs to have faith in their conclusions, like I have faith in mine. But I think the cumulative evidence from archaeology requires less faith from me to believe at least the core aspects of Jesus's life, ministry, resurrection, and its significance, than that required to continue to shrug off the stories as myth and wishful thinking. Did it happen? Is it true? While I might not be able to prove it deductively, there's evidence beyond a reasonable doubt that I think would stand in a court of law.[15]

In a sense, all of us are like I was in my early childhood: extremely near-sighted. We cannot see any significant distance into the past, and what is available to see at present is only a tiny fraction of that ancient world. What will we make of the data that we *do* have, and the trend that it shows?[16] What I have found is that through the eyeglasses of a relationship with Jesus, based solely on his sacrifice to cover my sins, I have a clear and robust perspective on the present, the past, and the future. So, does it matter? Most definitely.

Chapter 8

Why Science Was Stifled in China

Cynthia Cheung

"That's one small step for man, one giant leap for mankind!" I watched the Moon landing on television together with the young women in my dormitory at the University of California, Berkeley. It was my first summer as a college student in the United States. This human exploration milestone was especially captivating to me because I was an astronomy major. Still, I scarcely dared to dream that perhaps one day I could work in the space program and explore the universe, and maybe help others "go where no man has gone before."

Early Career Aspirations
I was born in Hong Kong and attended Christian schools built by missionaries. As a result, I have always believed Jesus as my Lord and Savior. My desire to be an astronomer was nurtured in my school days when there was a strong emphasis on studying science. The consensus was that, because China had isolated herself from the rest of the world for many centuries, she had fallen behind the West in science and technology. So, all students were encouraged to study science. I excelled in mathematics and science and was drawn to astronomy through a TV lecture and the *Star Trek* series. Though I lived in a small place, reading books lifted my imagination beyond my limited physical surroundings and let me explore many subjects, including outer space.

Early on, I started thinking about how science related to faith. Were science and faith decoupled, in conflict, or in harmony? I'll answer this question in two ways: (1) Would the pursuit of science lead me away from my faith? and (2) Would my Christian faith serve the culture in which I was raised? (I'll address this question a bit later.)

As a young academic, I prayed long and hard and asked the Lord for

guidance. I decided that God would show me the path through the college ap-
plication process. If I did not get admitted to a university that offered a good
astronomy program, then I might look to do something else.

Apologetics in College Days

I could not ask for a more interesting first-year American experience than
landing at UC Berkeley at the height of the student movement in the late 1960s,
with protests on campus almost every week. Walking through Sproul Plaza
and its many booths on the way to class, I felt like the apostle Paul perusing
the different altars when he paced the streets of Athens in the first century
(Acts 17:16–34). Among the booths that promoted different causes were ones
staffed by InterVarsity Christian Fellowship, Campus Crusade for Christ, and
the Berkeley Chinese Christian Fellowship (of which I was a member). My
involvement in those organizations was my first apologetics experience in the
US. I learned to engage students in dialogues about their faith at these booths.
I also met several scientists in the Jesus Movement[1] who came to Berkeley to
evangelize the hippies. These scientists had a definite influence on my pursuing
a scientific career as a Christian.[2]

Besides studying physics, astronomy, and mathematics, my university experience caused my faith in the God of the Bible to become firmly established.
The countercultural atmosphere (of flower children, opposition to the Vietnam
War, and recruitment into the Peace Corps) certainly challenged students to
question traditional values and beliefs and one's life purpose. When I observed
American young people delving into Eastern mysticism and transcendental
meditation, I experienced some sort of countercultural shock. I had been gradually adapting to Western ways, at least outwardly in my daily living, but now
I had to face daunting questions about my own identity: Am I abandoning my
thousands-years-old Chinese culture by studying in the West, or even worse,
by adopting a Western religion? Am I, like the hippies, participating in a countercultural movement, but of a different kind?

It was God's grace that put people in my path when I needed answers to
various questions. The Chinese Christian Fellowship was over 100 strong,
with students mostly from Hong Kong and Taiwan. (Mainland China was still
closed at that time.) We wrestled with our role as the first generation of Chinese
Christians and as future scientists in the Christian faith. What does it mean to
be Chinese, to be a Christian, and to be a scientist? Could we be true to ourselves, to our culture, and to God's calling? We concluded that every generation
has the responsibility to break away from the old, when necessary, and accept

the truth once it is known and move civilization to the next level.

Our fellowship counselor was Reverend Peter Yuen, a third-generation Chinese American and a Berkeley engineering graduate who dedicated his life to working among international students. He was a true gentleman—quiet but incredibly wise. He often gave this advice to seekers, "One cannot simply presume that God does not exist and dismiss him. You must be fair and give God at least a 50-50 chance. You could pray and ask God to show you that he exists." Many students prayed that prayer and came to faith.

Rev. Yuen also emphasized that if something were the truth, then one could throw anything at it and it would still stand. With that in mind, we invited science faculty who were Christians to come and speak to our group. They made it clear that science had its limitations and that scientists had also exercised faith in conducting science—faith in their axioms, presuppositions, and in the validity of their models.

Surprisingly, there were few questions about any conflicts between science and faith among non-Christian Chinese students. Rather, due to China's long history of war and poverty, the big question was why a good God would allow suffering. These students also thought that humans would no longer need God since science and technology could solve the world's problems. In other words, they put their faith in science—scientism—rather than in the God who created the universe and laid down the physical laws and intricate workings that scientists were just beginning to discover.

A Christian Astronomer

After Berkeley, I went on to pursue my PhD at the University of Maryland while working as a research assistant at NASA's Goddard Space Flight Center. There I began my affiliation with the space program in various roles until my retirement. The field of astronomy and astrophysics has advanced by leaps and bounds since my college days, with discoveries propelled by space missions and new ground-based telescopes. Astrophysics is no longer a descriptive science. It is now a mathematics-based discipline that uses accurate and high-precision data to determine the validity of theoretical models.

Thus, to answer one of my concerns as a young person with a passion for science, the pursuit of science did *not* lead me away from my faith. Rather, I continually see evidence of God's hand at work in the universe. The heavens truly display the glory of God (Psalm 19:1). At times I was so overwhelmed when looking at new results that I could only bow my head and worship God at my workstation. At about the time that Goddard Space Flight Center announced the precise measurements of 3°K cosmic background radiation (1990)—a telltale signature of the big bang—by the COBE satellite,[3] a colleague in computer science (a dear Christian brother) asked me, "What about the big bang?" He wondered how I could reconcile my belief in God with the new scientific discoveries. My response was, "Hallelujah! Because there was indeed a beginning!" (Genesis 1:1).

Why Science Did Not Develop in China

My scientific studies in the US had constantly confirmed my faith, but I kept thinking of my homeland. A nagging question kept coming to my mind. China had laid claims to many great innovations in the past, such as the compass, gunpowder, papermaking, and printing. Ancient Chinese astronomers made very detailed observations of solar and lunar eclipses, supernovae, comets, and meteor showers, and created star catalogues. But why had science and technology in China become stagnant for centuries? Why did it not continue to blossom? Why did the sixteenth-century Scientific Revolution take place in Europe but not in China?

I soon learned that I was not the first person to ask this question, which is known as "Needham's Grand Question." Joseph Needham (1900–1995) was a British biochemist, science historian, and sinologist. He initiated and authored a large part of a 7-volume monumental work, *Science and Civilisation in China*.[4] I can't give a comprehensive description of his work here, but I can relate my personal observations and struggles as one who was brought up in the Chinese culture and lived in its reality yet worked in a scientific field in the West. I see at least five reasons for why science did not flourish in China.

1. A Culture That Expects Conformity, Not Creativity

Chinese society (and other Asian societies) has been shaped by the teachings and the moral and ethical ideals of Confucius[5] for more than 2,000 years. Confucius was a sage born in 551 BC. He has been revered and even worshipped by some Chinese. His works *Four Books* and *Five Classics* are required studies for any educated person. I memorized parts of them in my Chinese literature classes in school. Confucian philosophy is so deeply ingrained in the culture that it is incorporated into the fabric of daily conversations and held as the governing moral principle for all human relationships. (My Chinese given name "ren-ai"[6] signifies the highest ideal in Confucianism.)

The Confucian "ten virtues" (benevolence, righteousness, courtesy, wisdom, faith, loyalty, filial piety, integrity, shame, and courage) have helped sustain a stable society for centuries, with adherents following a well-defined set of behavioral rules in the conduct of state, family, and education. These are excellent qualities to be attained by the "jun-zi" (virtuous man). At first glance, these virtues sound like the biblical fruits of the Spirit (Galatians 5:22–23). But in practice, this propensity to conformity does not encourage creativity or allow any deviation from the norm or the questioning of the status quo. Simply asking clarifying questions or voicing any skepticism about what is spoken by someone older or in a higher position (e.g., teachers, parents, the emperor) is considered disrespectful or even rebellious. It has a detrimental effect on the development of new ideas, not to mention scientific breakthroughs that require out-of-the-box thinking.

I came through such an education system, sitting in class as a good and obedient student through my elementary and secondary school years. I don't recall that I ever raised my hand to say anything unless I was asked by the teacher. So, when I first came to the US, I found it difficult to formulate questions or participate in a discussion because I had long suppressed my inquisitiveness and personal views. I realized that I had outgrown that mode when a NASA colleague approached me and "apologized" after a training session. He said that he had misjudged me as another "quiet Asian," until I spoke up and participated like any American. We had a good laugh about that!

Confucianism is a humanistic philosophy with no regard for revelation from a transcendent God who is the ultimate authority. Confucius himself stated that he would avoid broaching metaphysical or spiritual matters: "To be attentively respectful towards ghosts and spirits but keep them at a distance."[7]

So, how does one keep order in society? Respect the elderly and those in high positions and submit to them. This deference has resulted in the view that whoever/whatever is older or in a higher position is better, and most likely correct, and hence they become the (sometimes, ultimate) authority. The culture tends to look backward to something old, instead of forward to something new. I often hear the saying, "This has been done for several thousand years!" In a culture known for its culinary offerings, restaurants would advertise that they're using ancient authentic recipes, instead of *new* flavors created by their chefs! I don't deny that there's wisdom in ancient practices and the elderly have much to share from their life experiences. But this view has become a deterrent to societal improvements and refinements in working processes. Very quickly, science and technological development became stagnant.

Conversely, the Bible states that God does new things that are completely unexpected, like making a way in the wilderness and rivers in the desert (Isaiah

43:19). A quick word search through the Chinese Bible shows that there are over 200 occurrences of the word "new" or "renew"; for example, *new* heart, *new* life, *new* covenant, *new* song, and the *new* heaven and *new* earth. Christians who believe in the God who created the universe and sustains it have confidence that new things in his creation can be found. Scripture gives them the confidence to conduct scientific research, to do "search and discovery."

God asked people who trusted in him to leave their familiar surroundings (e.g., Abraham in the Old Testament and the disciples in the New Testament who were given the Great Commission) and go into the unknown where they would find blessings and bring salvation to the world. Many explorers came from the West (e.g., Columbus, Magellan, and Marco Polo) and early Christian missionaries traversed the treacherous Silk Road to reach China.[8] They expanded their knowledge of the world at the time and brought back exotic items from expeditions that enriched their society, stimulated cross-cultural exchanges, and encouraged further exploration.

I lived in the Washington, DC, suburbs for many years, where Columbus Day is celebrated annually. The first NASA Space Shuttle was named *Columbia* to signify the spirit of exploration. However, I would be hard-pressed to find explorers from China, except perhaps Admiral Zheng He[9] in the fourteenth century. His fleet might have reached the horn of Africa, but due to the displeasure of the Ming dynasty emperor, funding for his naval adventures dried up and the records of his accomplishments were mostly erased from official history.

In a Confucian society, historically there was little incentive to pursue a career in science and engineering. Social status was generally classified into nine echelons—with literary scholars on the top tiers, while people with science and engineering skills were relegated to the lower levels.[10] Even a medical doctor was ranked lower than a literary scholar. For more than 1,500 years, the court officials were scholars selected through an imperial examination system. The tests elevated only those who were conversant with the Confucian classics and philosophy and demonstrated their ability to compose prose and poems that carried the "Dao" (or "the way"). It was as if the college entrance examination (e.g., the Scholastic Aptitude Test) would only test aspiring students' understanding of Shakespeare but not their ability in mathematics.

2. A Worldview That Sees Nature as Virtual Reality
At the time of Confucius (known as the Warring States period), a plethora of philosophies and religions were at play, but they eventually got adopted,

assimilated, and integrated into the three main Chinese religions: Confucianism, Buddhism, and Taoism. Elements of other philosophies and mystical practices were retained within the three main religions. Thus, it's hard to find "pure" or original forms of these religions today. The three main religions are so deeply embedded in traditional thoughts and practices that some people consider them part of the Chinese culture and identity.

Confucianism, Buddhism,[11] and Taoism are nontheistic religions, meaning they do not worship a god. They coexisted for centuries because Confucianism, strictly speaking, is not a religion, but a humanistic philosophy that deals with the present life. Buddhism and Taoism, however, fill in the gap by addressing the spiritual realm: humanity's origin and future (life after death), suffering, and why the physical world is the way it is. For all three, their view of the physical world and man's relationship to nature differs significantly from the Christian worldview. Since natural science is the study of the physical world, these three views have strongly affected the development of science in China.

China's embrace of Buddhism could be partly explained by the inordinate amount of suffering she encountered. Except for a few dynasties, China's long history seemed to be marked by incessant wars (both internal and external) and natural calamities. Buddhism offered a way to break the cycle of suffering (life, aging, sickness, and death) via a path to enlightenment. On this worldview, suffering is caused by human emotions and desires for material things. The key is to empty oneself, become entirely detached from worldly things and relationships, or void all senses of self. This implies that the physical world (including the human body) is bad. One way to accomplish this detachment is to consider the material world not part of reality. In other words, one could deny its existence or believe that it's only a figment of one's imagination. To use a twenty-first century analogy, the physical world is virtual reality! It's obvious that science will not develop if the reality of the physical world can't be affirmed. What is there to study if the object doesn't exist?

The first book of the Bible tells us that God is the Creator of the physical universe. He preceded it, stands apart from it, and is above it. The creation is real and good. God delights in his finished work. Humans were created in the image of God, imbued with his Spirit, and were given the responsibility to manage the planet. Humans also stand apart and are a distinctly separate entity from the rest of the material world. Thus, we can study and examine the physical world with care, and the subject to be studied is real.

Both Buddhism and Taoism teach reincarnation and subscribe to the view that life/death is a cyclic progression. After death, a person returns to a different station in a future life, depending on one's good or bad deeds in the present life. Events in a person's present life are determined by karma (lit. "cause and effect" or deeds in the person's past life). This view leads to a fatalistic or pessimistic view of life since everything is predestined. Many people simply accept circumstances with resignation instead of seeking changes or finding solutions to create a better future—which is part of scientific research. This plight resonates with what the preacher said in Ecclesiastes, "Vanities of vanities! All is vanity. . . .What has been is what will be, and what has been done is what will be done, and there is nothing new under the sun" (Ecclesiastes 1:2, 9, ESV).

Chinese novels and movies are full of stories about escapism, such as the main characters checking out of life to enter a Buddhist monastery due to loss of love or other traumatic events. A remedy to bad karma might be doing good deeds so that one could live a better life, if not in the present, perhaps in the future. Note that the act is self-centered, not done out of love and care for others. There is no assured way out of the cycle except to devote oneself to meditation and possibly reach enlightenment.

It's a stark contrast to Jesus's teaching that he did not ask us to leave the world, but he sent us into the world (John 17:15–18). Christians are *in* the world but not *of* the world. "In this world you will have trouble. But take heart! I have overcome the world" (John 16:33). Reincarnation is also contrary to the scientific and biblical concept that time is a linear progression, with a definite beginning and end. God, the Creator of time, is an eternal loving Father who sent his son *into time* to break the cycle of sin and shame and offer people a new beginning.

3. A Spirituality That Seeks to Be One with Nature
Some Westerners think Eastern religious traditions have a higher regard for Earth's ecology because those traditions seek to be "one with nature." Confucianism, Buddhism, and Taoism all have the concept of "Tian" or "heaven" as the force that governs the universe and human affairs. The highest ideal is to reach the status that "heaven and man are one."[12]

One consequence of this Eastern view is that observations of nature have been used to deduce moral lessons for human conduct, but whatever happened in the heavens is quickly set aside. The motto of Tsinghua University, a top Chinese science-technical college, illustrates this well. Translated to English, the motto is "Self-Discipline and Social Commitment." It is distilled from the saying from the Taoist "Yi Jing" (*Book of Change*): "As the heavens maintain vigor through movement, the gentleman should constantly strive for self-perfection. As the Earth is vast, the gentleman should have the breadth of character to carry different circumstances."[13]

How did this view affect science? I'll use astronomy as an example. "Harmony between heaven and man" resulted in astronomers' work being used for divination. Court astronomers became imperial political advisers. They studied celestial phenomena and created detailed almanacs with accurate positions of the Sun, the Moon, and planets to predict the fate of the kingdom. The almanacs were mainly used as astrological charts, though they were also used as the farmer's almanac (to forecast seasonal weather, to determine the times for planting and harvesting, when to expect the first frost, etc.).

When I lived in Hong Kong, the farmer's almanac was sold only in Buddhist temples. Since people believed a person's destiny was tied to the celestial positions at one's birthdate, they would consult the almanacs for guidance to set the date for important events (e.g., weddings and funerals) and for matchmaking to determine compatibility between eligible marriage partners. Many of these practices remain in use in present-day Chinese communities. Other than for astrological uses, there was little incentive to do further research into *why* the stars and planets were arranged in such positions in the sky and *what* governed their orbital motions.

Before the fifteenth century, Europeans also comingled astronomy and astrology. But astronomy quickly became a scientific discipline after the Enlightenment in Christian Europe. Isaac Newton explored the *why* and *what* questions and came up with the universal law of gravitation, which eventually led to Einstein's theory of general relativity and big bang cosmology.

"One with nature" also means that everything must be arranged so that there is no conflict with nature. This means achieving a delicate balance between yin and yang (lit. shade and sun, essentially a balance between opposites, such as brightness and shade, hot and cold, etc.), and the five elements that make up nature: gold, wood, water, fire, and earth. Taoist values permeate Chinese medicine alternative treatments for various ailments. It is noteworthy that the yin-yang symbol is displayed prominently in many Chinese medicine practitioners' offices.

Sickness is due to the body going out of balance, thus regaining balance is the road to good health. For sure, many good dietary guidelines come from traditional Eastern cooking, including the consumption of more vegetables. But the modern concept of disease-causing germs and viruses is absent in Chinese medicine. Some practitioners claim to have found the panacea for some ailments, but those claims need to be met with skepticism and not accepted until they've been supported by scientific investigation. My frustration in caring for some of the sick in the Chinese community is that people would resist using proven Western medical treatments until sometimes it is too late.

The Taoist practice of feng shui (lit. "wind and water") is widespread in Hong Kong. A feng shui master would be consulted before one builds a residence so that it's perfectly aligned with nature, lest bad things happen to its occupants. This mentality has created a propensity for paranormal phenomena, even pseudoscience. One Christian friend commented, "Chinese may not believe in God, but they do believe in ghosts!" Experienced missionaries have warned people that the spiritual realm has a dark side.[14] After all, Satan could appear as an angel of light (2 Corinthians 11:14). When I took a class in martial arts, my tai chi master teacher (a Christian brother) told me to practice only the physical exercises of the discipline and not to touch transcendental meditation. Some of his colleagues had developed a psychosis as they were trapped by spiritual darkness when they sought to be "one with the force" to achieve a higher level of martial art.

4. Geographical Isolation That Leads to Ethnocentrism

China's isolation from the rest of the world can be partly understood from a geographical perspective. Cut off from the rest of Asia by a high mountain range (the Himalayas and the Pamirs Plateau) to the west and southwest, the forbidding Gobi Desert to the northwest, the frozen tundra of Siberia and Russia to the north, and a big ocean to the east, there was no easy way to gain access to the country. It was no surprise that China considered herself to be the Middle Kingdom, meaning the most advanced civilization in ethics, philosophy, and economics. China interacted only with people from a few surrounding tribes that eventually got assimilated. Even after the establishment of the Silk Road through the mountain passes that allowed travel to Europe, the flow of traffic was mostly one way. People had little interest in going beyond the natural barriers to see the rest of the world—that is, until Genghis Khan conquered China in the thirteenth century.

Then, in the nineteenth century, China was pressured to open trade with the West via foreign military power in the infamous Opium Wars. In 1842, Hong Kong was ceded to British sovereignty as the result of the unequal Treaty of Nanjing.[15] Thus, the modernization of China began when it started interacting with the West.

However, the tendency for ethnocentrism (the attitude that one's own group, ethnicity, or nationality is superior to others) remains to this date. I have detected this sentiment among some Chinese scholars who want the science and technology of the West but not their foreign God. It's natural to have pride in one's homeland. That's why we have Olympic Games and World Cups. But an air of superiority is contrary to the gospel message.

Thus, Christians should be ready to share our faith with gentleness and respect (1 Peter 3:15). An important message to convey to hearers of the gospel is that the God of the Bible is not a foreigner's God. This entails that we respect people's cultures and ethnic upbringing. When preaching the gospel of Jesus Christ that is meant for all people, Christians need to have a servant's attitude and separate the gospel from our own culture and denominational differences.

British missionary Hudson Taylor (1832–1905) established the China Inland Mission (CIM) in the mid-nineteenth century. He adopted Chinese attire to identify with the Chinese people and accepted missionaries across all denominations. Many were medical doctors and teachers who played a vital part in the modernization of China. They were transparent and let others see Christ incarnate in their lives. Until then, people might not have accepted scientific or technological advice, much less spiritual advice, until they knew Christians were authentic practitioners of their faith.

5. Faulty View of Human Origins

Because of the pride for their culture and homeland, many Chinese still hold to the concept that they are a unique people. Herein lies the resistance to the gospel. That is, a convert to Christianity may be looked upon as forsaking their own people and culture. In presenting evidence for resolving science and faith issues, I've deliberately tried to find references published by Chinese scientists in Chinese journals. And though my specialty is not in the biological sciences, I'll cite an example of receptivity to the Christian message concerning the origin of *homo sapiens sapiens*.

Due to political turmoil in the last 200 years, many Chinese are scattered overseas and have never seen their homeland. Members of this diaspora often have a particular interest in tracing their roots. Thus, Christians in the lands where the Chinese have settled can help them find their true root, God their creator, so that they can be rooted in eternity.

At one time some scientists claimed that Chinese people evolved from a separate branch of *Homo erectus* than Europeans and other peoples. The findings of the Peking man (a partial skull of *Homo erectus*)[16] in 1929 reinforced that belief. This discovery convinced many researchers that humans first evolved in Asia.

However, research by molecular anthropologists using DNA data have traced the origin of mankind back to a single male (Y-chromosomal Adam) strand and single female (Mitochondrial Eve) strand, located in East Africa.[17] This evidence demonstrates that the biblical creation story about the origin of humanity is correct. When I taught a science-and-faith Sunday school class, many students found this discovery helpful—especially when I described the work of a leading geneticist in China who reached the same conclusion.[18]

How to Bring Hope
Since the opening of China in the twentieth century, a significant number of Chinese have studied science in the West. Their experience leads them to ask the same question once asked by Joseph Needham and by me: Why did modern science develop in Europe rather than in China? This opens the opportunity for Christians in science to talk about the importance of Christianity to the development of science in the West. We can tell them that the scientific method originated from the Bible and that science could not have flourished if not for faith in the God who created the universe, stretched out the heavens, and set down its ordinances (Isaiah 44:24; Jeremiah 33:25).

Scientific evidence alone will not bring people to Christ, but the love shown by God's people will. Many Chinese scholars in the US and other countries have said that they walked into a church for the first time and immediately felt warmth and care like they've never felt before. The high moral and ethical standards in Chinese philosophies, traditions, and the conforming society do not address their personal concerns. Loneliness, fear, and uncertainties fill their hearts even though they are top students and scholars. They need to see authentic Christianity in the lives of people for them to trust the living Savior. Then the apologetics can follow to strengthen their faith.

Chapter 9

Reflections from a "Rocket Girl"

Leslie Wickman

My interest in science began developing when my father took my brothers and me outside on clear, starry nights—a somewhat rare occurrence in the Pacific Northwest where I grew up. Looking through his telescope at the Moon, stars, and planets sparked my early passion for astronomy and all things space related. As Christians, we had a shared understanding that God was the Creator of everything we saw, both on Earth and in the heavens. In addition, for as long as I can remember, I've felt close to God—so close that he was my best friend. In many ways, it seems that God has been preparing me for a career in science and faith for most of my life.

During my youth, my family lived in tiny towns and rural parts of Washington and Oregon. As a kid I was almost painfully shy and introverted. Give me a book or a cardboard box for a doll house or let me go play by myself on the beach or in the woods, and I'd be happy on my own for hours on end. From a very early age, I loved to read. Even after "lights out" I'd read with a flashlight under the covers. I'm convinced that I ruined my eyes in the process. I was the only one in my family who needed glasses before 40. Bookmobile day was one of my favorite days of the week!

Despite my shyness, God kept putting me on stage. From first grade on I was reciting Bible verses and singing solos in the school Christmas pageant. In later years, I was consistently elected to various leadership positions for school, athletics, and work organizations.

When I was seven years old, our pastor explained to my Sunday school class how to ask Jesus into our hearts to forgive our sins and bring us into a right relationship with God. I decided then that I wanted that for my life. Then, when I was nine, I went to a summer Bible camp. It was a memorable experience. I was crowned "Queen of the Camp" and carried around on a throne for memorizing the most Bible verses. But beyond the fun, something that has shaped my lifelong relationship with God happened during that week. Missionaries were invited to speak to the campers, and God laid on my heart that he was calling me to be a missionary in a far-off land. At that time, it seemed like the most frightening thing I could imagine. I went back to my cabin night after night and cried myself to sleep. Finally, I surrendered, and told God that I would do whatever he wanted me to do. Immediately I was overwhelmed with a profound sense of peace.

The main thing I retained from that experience was identifying with the story of God asking Abraham to sacrifice Isaac on the altar. It was a test of obedience no matter what, with complete trust in God's faithfulness to take care of the outcome. Ever since then, I've been careful to seek God's guidance at every fork in the road, watching for the providential opportunities he brings my way. And God has always been right there with me, faithfully leading me, protecting me, and encouraging me.

An Early Challenge to Science-Faith Compatibility

Even though I learned both at home and church that God was the source of every created thing, that idea would be challenged. I went to secular schools from grade school all the way through graduate school. Repeatedly, I heard in my science classes that the facts of nature have nothing to do with God or religion. A junior high biology teacher, a self-proclaimed atheist, would overtly tell students that we might as well leave our faith at the door because what he'd be talking about in class would almost certainly contradict what we heard at church. Even as a junior high student I remember thinking, "If God is who he claims to be as the Creator of the universe I'm studying, then how can the truth about God contradict the truth about his creation?" How could the theories of origins I was being taught integrate with my faith in the God of the Bible?

As a result, my personal quest to reconcile science and faith began when I was quite young. In fact, when I was about 12 years old, I remember reading through pages and pages of genealogies in the Old Testament, counting the generations and calculating how many years might have passed between Adam and Eve and the birth of Christ. I wondered if the biblical accounts were scientifically and historically accurate.

Also, because I couldn't completely reconcile the fully materialistic explanations of origins taught in school with the explanations given at church, to a certain extent I mentally compartmentalized my faith in God as the Creator apart from naturalistic theories of origins that I didn't fully understand. Not until after many more years of studious investigation did I really begin to see how my faith and science could fit together without conflict. And that, in fact, has become an enduring quest and passion.

Owning the Faith

From Washington State, my family moved to the North Coast of Oregon, and that was the beginning of my lifelong southern migration! After high school, I went to Willamette University in Salem, Oregon, for undergraduate studies, then to Stanford University in Northern California for graduate school.

All through grade school and high school I was very involved with church activities and was even asked to give the sermon at my church on "Senior Sunday." (Giving some insight into my independent nature, I delivered a sermon from the first chapter of James entitled "Faith Without Works Is Dead" at my Lutheran church!) But for the most part, up until college my faith was fairly internal for me. As many college students do, I went through a period of re-evaluating what I had learned and believed as a child. Over time, I realized that the idea of the God who created the universe loving us and sending his Son to us so that we could have a relationship with him made a lot of sense to me. This process resulted in my embracing the faith as my own, rather than something that other people had taught me and I had simply accepted. It also became clear to me that my faith in God needed to be expressed externally in ways that other people could see. Ever since then, it has been my strong desire to serve people for the Lord in whatever way he might call me.

Aerospace Career

When I graduated from high school, the impression of contradiction between science and Christianity still confused me. As an undergraduate student I avoided the conflict by majoring in political science. But the ever-persistent, always-pursuing "Hound of Heaven"[1] dragged me back to science and engineering in graduate school. Talk about doing it the hard way! Nevertheless, that experience has given me real empathy for students struggling with the STEMM fields. God never wastes any of our experiences in life.

In between my master's and doctoral programs, I started working at Lockheed Martin on both the Hubble Space Telescope and International Space Station programs. Before long I was designated as Lockheed's corporate astronaut. I went through literally hundreds of hours of simulations and training exercises. I earned my FAA pilot's license, got certified in high altitude physiology training in military hypobaric chambers, and went through parachute training. Getting my scuba certification allowed me to participate in neutral buoyancy (weightless environment) simulations in gigantic water tanks. I also participated in many flights aboard NASA's KC-135 research aircraft, which is affectionately referred to as the "vomit comet." I spent more than a decade doing this type of work and continue to do it on a part-time basis since I transitioned into a full-time career in academia.

While immersed in my aerospace career, I kept exploring the interfaces between science and faith through reading, research, and conversations with anyone who was remotely interested in the topic. I came to realize that most of the people I interacted with (whether at work, church, or in my social circles) hadn't thought too deeply about this subject. Typically, they assumed science and faith were either battling each other or had nothing to do with each other. Meanwhile, my own position was gravitating toward one of fully integrating the best practices of biblical interpretation with well-researched science.

Faith-Affirming Scientific Evidences
There are several strong threads of evidence from my scientific research that convince me to trust God's faithful revelation of himself in nature.[2] Among the most powerful threads is the observation that life-friendly planets seem rare, certainly in our solar system and maybe even in our galaxy and beyond. Producing the perfect conditions for life requires meeting many exacting parameters. The odds of getting all requirements within the narrow range of feasible values is, by most standards, statistically impossible. These finely tuned parameters include things like the very laws of physics that control the interactions of space, time, matter, and energy in our universe.

Strong Nuclear Force
For example, the strong nuclear force—which holds the neutrons and protons together in the nucleus of an atom—is the strongest force in the universe. Other forces, like gravity and electromagnetic attraction and repulsion, taper off gradually with distance. But the strong nuclear force goes from super strong to nothing at a very precise distance that extends to the average diameter of the nucleus of an atom (10^{-15}m). Any nucleus larger than that distance will be unstable and subject to radioactive decay. The greater the number of positively charged protons there are in the nucleus of an atom, the more neutrons are needed to overcome the repulsive electromagnetic force between protons with

the strong nuclear force that acts between all of the nucleons. There are also radioactive isotopes of low atomic number elements (including hydrogen and helium). However, technetium, with an atomic number of 43, is the element with the lowest atomic number for which all isotopes are radioactive.

So, if the strong nuclear force did not exist—or were weaker, or only good for shorter distances, or if the electromagnetic repulsive force were stronger— then our universe could consist only of the lower atomic number elements. In the extreme case, hydrogen would be the only element that could form. On the flip side, if the strong nuclear force were stronger or good for longer distances—or if the electromagnetic repulsive force were weaker or did not exist—then large, heavy atomic nuclei would dominate our universe. In the most extreme case, *all* the nucleons in the universe would be bonded together in one gigantic mass. Either way, the life-friendly universe we know and love would not be possible!

Universe's Features
Another set of significant finely tuned parameters in the universe is (1) its uneven distribution of energy and matter; (2) its mass density; (3) its expansion rate; and (4) the strength of the universal gravitational constant. If any of the values for these parameters were just a tiny bit larger or smaller, then the universe would have either expanded too quickly for matter to coalesce to form galaxies, stars, and planets, or conversely, the universe would have collapsed back into itself before any of these objects could have formed.

Earth's Features
Moving from the universal scale to the planetary level, we often read about the ongoing search for extrasolar planets. Of particular interest is whether a planet is in its host star's habitable zone: the distance that provides the right temperature range for liquid water (essential for biological life). But that one criterion is just the beginning! There are many other attributes that go into building a life-friendly planet. For example, various combinations of a planet's climate, size, mass, density, structure, composition, and rotation rate determine how strong its gravity is, whether it has a protective magnetic field (and if so, how strong it is), how much and what kind of atmosphere it can retain, how thick its crust is (or if it even has a solid crust), how much seismic and volcanic activity it has, and the list goes on!

Water's Features
One specific example of fine-tuning relates to water. I like to call water the "miracle compound." The water molecule, H_2O, weighs in at 18 grams per mole. Water—in all three physical phases—is essential for life. Earth's atmospheric pressure and temperature are all just right for supporting the cleansing, purifying, life-giving water cycle. Intriguingly, Earth's gravity is strong enough to hold on to ample amounts of biophilic (life-friendly) water vapor in the atmosphere at 18 grams per mole, but not strong enough to keep large amounts of toxic methane (CH_4) at 16 grams per mole, or ammonia (NH_3) at 17 grams per mole.

Moreover, water is rare among abundant compounds in that it expands when freezing, rather than contracting. This is fortunate—or rather, providential! It means that water ice is less dense than liquid water and it floats, insulating the liquid beneath. If water behaved more typically and contracted when freezing (sinking rather than floating), then bodies of water exposed to cold winter temperatures would freeze solid in the winter, and possibly not warm up enough to melt thoroughly in the warmer seasons.

Quantum Uncertainty
Quantum uncertainty suggests a probabilistic rather than a deterministic universe. String theory suggests extra dimensions of space and time. The latest scientific ideas about these theories corroborate a biblical worldview. As we learn more about the universe we inhabit, we discover a more complete understanding of what we think of as physical versus spiritual, and natural versus miraculous. My faith is further confirmed in knowing that at both the smallest and the largest scales of creation there is evidence of the Creator.

Science Reveals a Creator's Care
In summary, conditions on Earth are not just barely survivable, as if the Creator were doing some kind of biology experiment to determine the minimum conditions for life! Conditions are ideal for intelligent creatures to *thrive*. This just-right environment speaks to me of the loving God we read about in Scripture.

All my research on the amazing intricacies and synergies that God designed into both the universe at large and our home planet inspires me to be faithful to God's command for us to be stewards of the world he so carefully created for us and called "good" over and over in Genesis 1. It also brings deeper meaning to the beloved passage in John 3:16: "For God so loved the world [*cosmos* in the original Greek] that he gave his one and only Son."

A Passion to Help People See the Harmony
My transition from full-time aerospace work to full-time academic work came when I realized that God had provided me with some unique experiences and resources for sharing key concepts connecting science and Christianity. I went into academia largely because of my passion to help students wrestle through the perceived conflicts of science and faith that I had wrestled with in secular universities and workplaces. I want to help people—especially students—understand that the competition between science and faith is an illusion. The truth about nature and about nature's Creator must complement (rather than contradict) each other since the Creator-God is One. And as such, he is the source of all truth.

Students are still discovering their paths in life. It's important for them to realize that they can be both faithful Christians as well as top-notch scientists. Furthermore, I believe that everyone is uniquely designed, and that God has a loving plan to give each one a future and a hope (Jeremiah 29:11). I desire to help students gain an appreciation for the unique individual that God has designed them to be and to encourage each one to develop their full potential.

Earlier I said that God had called me at a young age to be a missionary in a far-off land. Here's an example of how that calling has been realized. After returning to academia as a physics and engineering professor, I was teaching a college astronomy course when news broke about possible evidence for Einstein's predicted gravity waves. CNN asked me to write an article for their website.[3] In it, I mentioned that science is a tool we can use to uncover the wonders of God's creation. The public reaction was nothing short of astounding: the article went viral, with over half a million views in less than a week and 70,000 shares on Facebook. It was in the top 5 worldwide most-shared news stories on social media! On top of that, I received tons of emails. The direct messages were mostly positive, but peppered with a little hate mail, just to keep things interesting. One person even got my personal cell phone number and called to give me a piece of his mind on the matter. That experience made me realize just how polarized people can be on the topic of science and religion.

In the wake of all the publicity from that CNN article, I was offered a book contract for *God of the Big Bang: How Modern Science Affirms the Creator*, which led to doing a TEDx Talk.[4] From there I began getting a lot of requests to speak at churches, schools, and other organizations about how science and faith go together.

In my science and faith lectures I always point out how God reveals himself in the two books of creation and Scripture, and how God invites us to engage with him in figuring out how his two books fit together. As Proverbs 25:2 tells us, "It is the glory of God to conceal a matter; to search out a matter is the glory of kings." This is a working partnership between God and humans. If God truly is the Creator we believe him to be, then he will reveal himself through what he's created. Science is the tool we can use to uncover those wonders. When both books (nature and Scripture) are properly understood, they coexist peacefully. Much of the most compelling evidence for that science-faith compatibility is summarized in this list:

- There is something rather than nothing.
- The universe had a beginning, therefore a cause, which the cosmological argument claims.
- The universe has orderly physical laws and extensive fine-tuning consistent with the teleological argument.
- Quantum mechanics allows for the improbable without violating physical laws.
- Modern science was firmly established by devout Christians who believed that Scripture encourages its readers to study science.
- Many famous top-notch scientists profess faith because of, not in spite of, evidence from nature.

A Peaceful Path Forward
Many resources exist for exploring this summarized list in detail. I've mentioned my book and I also recommend Reasons to Believe (reasons.org). In addition to these resources, I make a habit of leaving every reader, audience, or classroom with some tips for irenic, fruitful science-and-faith conversations:

- Approach the dialogue with humility and grace, for as St. Paul writes in 1 Corinthians 13:12: "For now we see only a reflection as in a mirror."
- Do your best to understand the spectrum of positions and arguments on all sides.
- Realize that most people are on a journey and still figuring out where they stand.
- Recognize that these issues are not essential to your faith or your salvation.
- Learn to live with tension; it's OK not to have it all figured out yet.
- Don't let the arguments or disagreements upset you.

- Finally, heed seventeenth-century Lutheran theologian Rupertus Meldinius's advice: "In essentials, unity. In non-essentials, liberty. In all things, charity."[5]

I still feel that the Lord is my best friend, and he guides and directs my paths by opening opportunities for me to serve him in ways I could never have planned. Every time he does this, it makes me trust him more and shows that his plans are far greater than my plans. So far this has included teaching and counseling young people, building houses for people in Mexico, playing volleyball in South Africa and South America with Athletes in Action, playing women's professional football (and serving as captain and chaplain) with the California Quake, singing in Southeast Asia and Northern Ireland with the Witness Ministry Team, and working in the US Space Program.

My education, training, and experience as a "rocket girl" have led to countless opportunities to demonstrate the harmonious relationship between science and faith. The life experiences God has brought me far exceed any dreams I've dared over the years, and I can't wait to see what he has for me next!

Chapter 10

Is There Compatibility Between the Big Bang and the Bible?

Michael G. Strauss

In the fifth year of my postdoctoral appointment with the University of Massachusetts, my professional career was thrown into chaos. I was doing research in experimental particle physics, studying the fundamental particles and forces that make up the universe, at what is now the SLAC National Accelerator Laboratory. I anticipated finishing my postdoctoral research in about a year and then becoming a faculty member at a research university. It was the early 1990s and the Superconducting Super Collider (SSC) was being built in Texas. The SSC was to be the world's largest and most powerful proton accelerator. I expected that my future research would be conducted at the SSC and that my career path would be linked with its success. Then, in October 1993, the United States Congress cancelled the SCC construction.

With the closure of the lab, almost all the SSC laboratory's senior physicists were unemployed and looking for work. Normally, in experimental particle physics there are about a dozen faculty positions at US research universities open per year and a few dozen postdocs applying for those jobs. But in 1993 and 1994, the number of job applications swelled to almost 200 people vying for a handful of positions. My goal of becoming a faculty member at a research university seemed unattainable.

A Search for Truth
But, as with many incidents in life, those seemingly disastrous circumstances led to new personal awareness and growth. I now faced the prospect of an unforeseen career transition and I had no idea what I wanted to do with my life. My wife encouraged me to spend a weekend in a hotel, away from her

and our two young children, to consider seriously what career options I might want to pursue. I purchased a copy of the book *What Color Is Your Parachute? A Practical Guide for Job Hunters and Career-Changers* by Richard Bolles and settled into my room. The book prompted me to evaluate my primary occupational goals and motivation, which I had not done before. With some reflection, I realized that I was motivated to "discover and understand truth."

Although this insight may not seem profound, it was a revelation to me that this desire to know the truth had been (and continues to be) a unifying theme in my approach to life and the decisions I have made. For instance, one of the reasons I had gone to graduate school and pursued a PhD in physics in the first place was that I saw science as an avenue for uncovering truth about the way the universe operated. My ongoing quest to discover truth about the universe has led to research in particle physics—specifically, in measuring the properties of the Higgs boson, which will help us understand outstanding mysteries of the universe. I also love teaching physics. It's gratifying to see university students learn how the universe works.

In science, and in other arenas of life, there are often competing theories. When evaluating various ideas the most important question I ask is, "Which alternative is most likely to be true?" This guiding question has helped me address issues like the alleged war between science and faith. I've found that following a few basic principles helps me answer this question. First, I must really understand the various viewpoints. Our social media culture often presents us with confirmation bias, which affirms only information we already accept while ignoring conflicting facts. Any search for truth requires that differing ideas be fully examined and understood. Second, any valid conclusion must explain all the data and facts with minimal contradiction.

Sometimes the graduate students I work with will show me graphs that they've produced from our data. Although each graduate student may believe that his or her graph portrays the experimental results accurately, I might notice that two different graphs display inconsistent results. Therefore, either the graphs have been made incorrectly or they show a new scientific discovery. The latter is much more exciting, but the former is usually the case.

Because truth is important to me, I tend to ask for evidence to support everything I believe. Being skeptical seems to be a consistent trait of most scientists. Because I value truth and will draw conclusions based on data, I have changed my beliefs or views about many different topics throughout my life.

Valuing truth, being skeptical, investigating different viewpoints, incorporating all the data into conclusions, and changing beliefs may be traits expected of a scientist. But I'm not only a scientist. I'm also a Christian. And when I approach the things I believe about Christianity, I don't discard these important principles. I embrace them. After all, my core personal value is to discover what is true, even in the realm of religious ideas.

Apparent Conflict over the Universe's Age

I grew up in a Christian environment. My father was the pastor of a church. I was taught that the Bible was God's Word and that the various stories in the Bible were true. My elementary school years were spent in Huntsville, Alabama, where the first stage of the Saturn V rocket that sent humans to the Moon was assembled and tested. The space program, science, and technology fascinated my young mind. This interest laid the foundation for me to become a physicist. As a child I saw no conflict between the world of science and the biblical world of Christianity.

However, in high school I began to realize that many scientists and Christians saw a huge conflict between the findings of science and biblical writings. This conflict seemed to be most prevalent when comparing the scientific explanation for the origin and history of the universe with the biblical creation account. My early explorations into this topic also seemed to reveal conflict. Most of the Christian authors I read at the time claimed that Genesis taught that the universe was only about 6,000 years old and that the scientific narrative of a 14-billion-year-old universe with a big bang origin misinterpreted the data due to improper presuppositions. Some Christians claimed that God had created the universe with only the appearance of billions of years of age. That is, God created humans and the universe in a mature state, so that when all things came into existence, they appeared older than they really are.

I understood the primary motivation for such beliefs from a Christian worldview. Christians believe that the Bible is God's Word and, therefore, accurate and true.[1] Some Christians thought that they were upholding the integrity of their sacred texts by proposing such ideas (apparent age) about the age of the universe. Of course, there are theological problems with the appearance of age view. If a god made the universe look like it had an ancient history when it did not, then that god would be deceptive. The Christian God does not deceive.

As a teenager and young adult, I was somewhat concerned about this apparent discrepancy between science and the Bible, but I did not do a thorough investigation for myself. I did believe that the Christian God speaks truth. Furthermore, I believed this God, as creator of the universe and author of the Bible, had revealed a consistent picture of himself in the physical universe and in Scripture.

However, as a graduate student studying physics at UCLA, I decided that if I was going to be a career physicist and continue to hold my Christian beliefs, then I had to determine what was true about the scientific and biblical record of the origin and history of the universe. I knew I would have to investigate various viewpoints and see which ideas best explained the data and were cohesive and objectively defensible.

Personal Investigation and Testing

It was about that time that I encountered writings by Hugh Ross,[2] an astrophysicist who was a Christian. He claimed that the account of creation in Genesis was in perfect agreement with what scientists had discovered about the origin of the universe in the big bang—an idea often referred to as "old-earth creation" (OEC). This proposal was quite different from the Christian writings I had previously read that claimed the Bible described "young-earth creation" (YEC). Ross and others who held to OEC claimed that a proper understanding of the original text of the creation account, written in the ancient Hebrew language, did not describe a 6,000-year-old universe, but *did* accurately describe what was known about the origin and development of our 14-billion-year-old universe and our 4.5-billion-year-old Earth.

I decided it was time I personally investigated what the Bible said about the origin and development of our universe rather than simply rely on what others had told me. Because the Genesis story of creation was written in ancient Hebrew, it seemed that a proper understanding of the text and its nuances would require some knowledge of the ancient Hebrew language and culture. Anyone who speaks multiple languages knows that certain subtleties and nuances may be difficult to convey accurately from one language to another.

Since I don't have a working knowledge of ancient Hebrew, I followed the next best path. My father's home office had thousands of volumes of theological texts. Many of them were unbiased reference works designed to help an English-speaking person understand the original biblical languages. These reference books included Hebrew-English interlinear writings with the Hebrew words directly linked to English translations, Hebrew-English dictionaries and concordances, Hebrew commentaries, and writings by scholars who were experts in the ancient Hebrew language and culture.

I then gathered some of the best books written by those who advocated a YEC view and those who advocated an OEC view. Both claimed that their understanding represented the biblical text most accurately. As I compared the claims of the YEC writers and the OEC writers to the unbiased reference works, in every case the OEC interpretation aligned more closely with the unbiased reference works. For instance, when the YEC writer would assert that a certain Hebrew word had a particular nuance and the OEC writer would assert a different nuance, the unbiased reference would support the OEC claim. This was an amazing revelation for me. The idea that is held by some Christians and by many nonreligious people that the Bible teaches the universe is only a few thousand years old, is just not true. Such a viewpoint shows a poor understanding of what the original text says.

My investigation of the biblical text began over 30 years ago. Since then, I have continually studied to try to understand the Bible's meaning, and to determine if it harmonizes naturally with our current scientific knowledge about the origin and development of the universe. I've found, quite remarkably, that the text matches exactly what science has discovered. This, to me, gives strong support of the divine nature of the biblical text. Somehow, a writer living at least 3,200 years ago wrote an account of the origin and development of the universe that aligns perfectly with twenty-first-century scientific discoveries.

If the account of creation in Genesis is true and accurate, then the veracity of the rest of the Bible seems worth investigating as well. So, let's take a journey through the Genesis account of creation, observe what it says and what it doesn't say, and compare it to our scientific knowledge about the 14-billion-year history of the universe.

Our Universe's Beginning

The first verse of the Bible, Genesis 1:1, says, "In the beginning, God created the heavens and the earth." The phrase "the heavens and the earth" refers to the entire cosmos because ancient Hebrew has no word for "universe." Although most scientists prior to the twentieth century assumed that the universe did not have a beginning—at least until Edwin Hubble's 1929 discovery that the universe was expanding[3]—the Bible declares that it had a beginning. Scientists technically don't know what happened in the first 10^{-30} seconds (an infinitesimal moment) or so of the universe's history. But all observations—including the expansion of the universe and the cosmic microwave background radiation[4]—and all theoretical calculations based on known physics—including general relativity as expanded by George Ellis, Stephen Hawking, and Roger Penrose,[5] and the Borde, Guth, Vilenkin theorem[6]—indicate the universe had an actual beginning as proclaimed in the Bible's first verse.

Planet Earth formed about 9 billion years after the beginning of the universe. Consequently, Genesis 1:1 comprises about 9 billion years of time.

Because the Bible primarily tells the story of God's interaction with humans, it's not surprising that the first 9 billion years of history before the formation of the planet that humans will inhabit are contained in one sentence. Additionally, I don't think that 9 billion years is a very long time to a timeless God.

The next sentence in the Bible may be the most important in the whole creation account for it sets the stage for the rest of the story. It says, "Now the earth was formless and empty, darkness was over the surface of the deep, and the Spirit of God was hovering over the waters" (Genesis 1:2).

Earth's Initial Conditions

There are two very important things to observe in Genesis 1:2. First, it describes the conditions on the primordial earth. The earth was formless (incapable of supporting life), empty (devoid of any life), dark (on the surface of the earth), and mostly water. This condition of being formless, empty, dark, and watery is identical to our current scientific understanding of the conditions on the early earth. Obviously, it was devoid of life and incapable of supporting life.

Additionally, it was dark on its surface because Earth had a very thick and dense atmosphere which, along with dust and debris in the atmosphere and in outer space, blocked out any light from celestial objects. The Sun, Moon, and stars would have already existed as their creation is described in the previous verse. But neither they nor their light were observable from Earth's surface due to the opaque atmosphere. If the Bible had said that it was dark *everywhere* when Earth was first formed, then it would have been wrong. Since it only says that it was dark on the planet's surface, then it is accurate. Finally, we see that the Spirit of God was "hovering over the waters," implying that the surface was mostly water and not dry land. Scientific investigations have confirmed that the primordial earth was almost all water with little land.[7] When the author of Genesis, writing more than 3,200 years ago, described the initial conditions on Earth as formless, empty, dark, and watery, he got it correct.

The Storyteller's Perspective

The second important thing to observe in Genesis 1:2 is the perspective from which the rest of the story is to be told. Every story is told from some particular perspective. The story in Genesis 1 is told from God's perspective since he is the one doing the creating. We are informed that the Spirit of God is hovering near the surface of the earth. In other words, the account of creation is not told from some viewpoint outside the universe looking in, or even from somewhere in cosmic outer space, but from the perspective of one near Earth's surface.

This is a crucial point! If one tries to understand the order of events described throughout the rest of the chapter from any perspective other than the one clearly indicated, then the description is scientific nonsense. But when viewed from a perspective on Earth's surface, the subsequent narrative correlates exactly with known science.

With Earth's initial conditions defined and the perspective of the story stated clearly, the rest of Genesis 1 describes the transition of Earth from formless and lifeless to a planet full of various life-forms over a period of six "days." The first three days address the transformation of Earth from formless to life-supporting, and the second three days address the development of life on Earth. The six "days" of creation, then, describe our planet's 4.5-billion-year history.

How Long Is a Day?

There's a great deal of misinformation regarding the length of time referred to in Genesis 1 because of the use of the Hebrew word *yôm*, which is translated "day." It would be a mistake to assume that every use of the English word "day" means a period of 24 hours. If I say, "In Lincoln's day the North and the South fought a costly civil war," I'm using "day" to refer to an extended period of time in the past. If I say, "It's a beautiful day," the usage suggests that "day" refers to a particular moment in time. In Oklahoma where I live, we might have had a thunderstorm pass through 10 minutes prior, but if the Sun is out and the sky is blue right now, it is a beautiful "day." Context reveals the meaning of a word. Like the English word "day," the Hebrew word *yôm* can mean a 24-hour cycle, the daylight hours, or an extended period of time, among other nuances.[8]

Only the literary context will communicate the meaning of the word *yôm* regarding the six days described in Genesis 1. Scholar Gleason Archer, who was fluent in Hebrew and over 30 related languages wrote, "On the basis of internal evidence, it is this writer's conviction that *yôm* in Genesis 1 *could not* have been intended by the Hebrew author to mean a literal twenty-four-hour day" (emphasis added).[9] According to Archer, the text of the Hebrew Bible itself indicates that the creation days are not 24 hours. The literary context indicates that each creation day may be best understood as a long period of time.

Ancient Hebrew has about 3,000 root words and a total of less than 9,000 words.[10] It has no word that means "a long but finite period of time" like *era*, *epoch*, and *age*.[11] (Modern English has about a million words, with over 160,000 found in a typical college dictionary.) If the writer of Genesis wanted to describe the history and development of the earth through various epochs, the best word he could use would be *yôm*. Consequently, *yôm* in Genesis could very well be translated as "era" rather than "day," and that may be the best translation. With that in mind, let's explore the sequence of events described in Genesis and compare that sequence with what we know from scientific inquiry.

Creation Day 1

On the first day God said, "Let there be light," and light appeared on Earth's surface. The text implies that nothing new was created on this day. The celestial objects had existed in outer space prior to day 1. Instead, the opaque sky became translucent to allow light to appear for the first time on the planet's surface. Now daytime could be distinguished from nighttime, but the Sun, Moon, and stars were still obscured, much like what we observe when the sky is cloudy. Scientifically, this atmospheric transition would be one of the first events observed on the surface of the earth.

Creation Day 2

On the second day God made an expanse that separated the waters above from the waters below. This event describes the beginning of a water cycle. The waters above are the clouds, and the waters below are the oceans, and the expanse is the sky in between. In *The Expositor's Bible Commentary* on Genesis, J. H. Sailhamer affirms that these verses in Genesis are describing the ordinary oceans, sky, and clouds we are familiar with, which contrasts with many ancient cultures that viewed the heavens as some solid hemisphere above the earth. Sailhamer writes,

> It would be unlikely that the narrative would have in view here a "solid partition or vault that separates the earth from the waters above" (Westermann, p. 116). It appears more likely that the narrative has in view something within the everyday experience of the natural world, in a general way, that place where the birds fly and where God placed the lights of heaven (cf. v. 14). In English the word "sky" appears to cover this sense well. The "waters above" the sky is likely a reference to the clouds.[12]

Creation Day 3

On the third day there were two creative acts. First, the waters were "gathered to one place" to create the oceans and "dry land" appeared. The planet was transformed from a watery world to one that had a significant amount of land. Scientists now know that this change occurred about 3 billion years ago when tectonic activity began. In a very short geological period, the continents arose from the ocean and the land appeared.[13]

The second act of creation on the third day was the creation of plants and vegetation. The terminology used, "Let the land produce vegetation," suggests a natural process in which plants are produced. The Hebrew word for vegetation (*deshe*), though often used for grass, can also be used of just about any plant that photosynthesizes, including microscopic algae. Interestingly, microscopic algae first appeared on Earth about 3 billion years ago, at the same time that continents first formed, both on the third "day."

Creation Day 4

On the fourth day, the Sun, Moon, and stars appeared in the sky as the atmosphere thinned enough to finally make the celestial objects observable from the earth's surface. Many skeptics have dismissed the biblical account of creation because they claim the text says that the celestial objects were created here on the fourth day after Earth was created, which is scientifically ludicrous. But this uniform assertion overlooks the fact that the account of creation is being told from a perspective on the earth's surface. The text doesn't say the objects were *created* on the fourth day, only that they *appear* on the fourth day and are given significance. Ancient Hebrew verbs do not have a conventional past or present tense, but instead have a form that means the action is completed and a form that means the action is in progress. The verb tense in verse 16 that says "God made two great lights" indicates completed action, namely that God had completed making the Sun and Moon sometime in the past, not necessarily on the fourth day. The verb could be translated as, "God had made the two great lights."

Sailhamer affirms that the events of day 4 do not describe the actual creation of the Sun, Moon, and stars. He writes, "It suggests that the author did not understand his account of the fourth day as an account of the creation of the lights; but, on the contrary, the narrative assumes that the heavenly lights have been created already 'in the beginning.'"[14] Again, we see that this description aligns with what we know from scientific inquiry. That is, from a perspective on Earth's surface, the Sun and the Moon are "two great lights" and their appearance in the sky occurred only after the planetary atmosphere had become transparent about 2.4 billion years ago when photosynthesizing microbes had produced enough oxygen that accumulated in the atmosphere.

Creation Day 5

On the fifth day God created many of the animals that fill the sea and the air. The original Hebrew language indicates two classes of animals that were created on this day: *sherets* and *nephesh*. The Hebrew word *sherets* can refer to any type of animal that swarms, teems, or creeps. This meaning encompasses a variety of fish and birds in this context. The Hebrew word *nephesh* can mean any living thing, but it also has a more focused meaning of "soulish" animals. The word *nephesh* itself comes from the Hebrew word that means "breath" and usually refers to soul, desire, or spirit.[15]

Soulish animals would then include only animals like mammals or domesticated birds. Skeptics will argue that the order of animal creation in Genesis doesn't follow the order we know from scientific investigation. However, using the more focused understanding of *nephesh*, the second class of animals mentioned as being created on the fifth day are those animals that we would think of as soulish, including sea mammals and the types of birds humans might keep as pets.

Creation Day 6

On the sixth day of creation two more classes of animals were created. First, God created more *nephesh* creatures (not *sherets*) that live on land. That is, the land animals mentioned as being created late in Earth's history are advanced mammals, in agreement with the paleontological record. To emphasize this point, the author specifically mentions the large mammals that interact with humans, such as cattle and other beasts of the field. From the nouns used by the author, it's clear that the accounts of days 5 and 6 are describing the creation of animals most important to human existence, rather than the creation of *all* animal life throughout Earth's history.

In the final and ultimate creative act, God made humans—beings who are not just physical and soulish but also spiritual. For the first time on planet Earth, one of its inhabitants was a spiritual being, made in the image of God. Of course, this creative act also agrees with the fossil record, which indicates humans are the most recent and most sophisticated creatures to inhabit Earth.

Science and the Bible in Harmony

To summarize God's creative acts over the six days in Genesis 1, we see 10 events. It is remarkable that, when we assume the proper perspective from the surface of the earth, the order of these events as described in the Bible is exactly the same as that observed by science:

1. The universe has a beginning and eventually the earth is created. The primordial earth is formless, empty, dark, and watery on its surface. (~4.6 Gya)
2. Light appears on the surface of the earth as the atmosphere thins and the planetary and interplanetary dust clears. (~4.2 Gya)
3. The water cycle begins. (~3.8 Gya)
4. The continents form. (~3.0 Gya)
5. Microbial algae and other plants emerge. (First cyanobacteria ~2.7 Gya)
6. The atmosphere becomes transparent, making the Sun, Moon, and stars visible. (~2.4 Gya)
7. Fish, sea mammals, and birds fill the sea and air. (Fish ~530 Mya, birds ~100 Mya)
8. Higher mammalian life-forms fill the sea. (~40 Mya)
9. Large mammals that will interact with humans fill the land. (Modern mammals ~10 Mya)
10. Humans are created. (~.1 Mya)

The details of creation written in the Bible more than 3,000 years ago agree with the modern scientific understanding of the big bang and the subsequent development of the cosmos over 14 billion years. When the record of nature and the words of the Bible are understood properly, there's no conflict. Instead, there's complete harmony between science and Scripture.

Arno Penzias, the Nobel Prize–winning physicist who discovered the cosmic microwave background radiation was quoted as saying, "The best data we have are exactly what I would have predicted, had I nothing to go on but the five books of Moses, the Psalms, the Bible as a whole."[16] As a person who is motivated by discovering truth, I find an amazing correlation between the biblical record and the scientific conclusions regarding the origin and history of our universe. This compatibility seems to indicate that truth, even regarding scientific phenomena, is found in the Bible. For me, the many truths found in the Bible have been both life-giving and life-transforming.

Chapter 11

Mission Accomplished

David Rogstad

This chapter was written before the author's death and was edited posthumously. This final version has been reviewed by David Rogstad's wife and children to confirm its faithfulness to the original message.

If a person could become a Christian simply by growing up among Christians, that would be me. You might have called me a cultural Christian, if there really were such a thing. I fit in comfortably with my family and church friends, but I didn't really take my faith seriously at a deep personal level until my graduate school years. A vitally important discovery awakened me—a personal one, not a scientific one.

My parents were immigrants from Norway, and I was the youngest of their five children. Dad was a skilled cabinetmaker whose example of being well-read and self-taught in the sciences inspired my brother, then me. I remember poring over my brother's science and engineering textbooks, trying to work out what they meant. My curiosity eventually led me to Caltech to study engineering. After completing a year in the engineering program there, I switched to physics, which I found even more fascinating.

I didn't participate much in Caltech student life, but I do have fond memories of solving puzzles and intricate math problems with some friends during a few boring moments in class. During my undergraduate years, I lived in the back bedroom of my parents' home studying, listening to music, and building my own hi-fi system. Every evening before going to bed, I would play Frédéric Chopin's Piano Concerto No. 1, loudly, to help me unwind.

A Guilty Conscience

Although my parents had always emphasized the importance of honesty and integrity, I lacked their strength of conviction. In fact, I sometimes ignored that conviction. In the summers between my undergraduate years, I worked at Boeing in Seattle, thanks to my brother's connection there. I lived with him and his family and worked on 707 jets. During my final summer at Boeing, I came across some of the parts I needed for the hi-fi system I was putting together. They were in an electronics storeroom, and I justified my taking them by telling myself that these little parts would be an insignificant loss to such a huge company.

Later that year, I happened to visit a Sunday morning class at the church my parents attended. The group was studying the book of Ephesians. My visit made me painfully aware of how little I really knew about the Bible, despite my upbringing. Friends I had known for years were actively and insightfully participating in the discussion, and I had nothing to add. My pride was stung. A surge of envy swept over me. I had always thought of myself as smarter than these people, but here they had surpassed me.

That afternoon I asked my parents if they'd consider buying me a good study Bible as my college graduation gift. My big plan was to read it over the summer before beginning grad school. My parents were delighted by the request and presented me with a crisp new Scofield Bible, which I proudly added to my bookshelf.

During my first year in Caltech's graduate physics program, the course work seemed to come easily, and I spent many evenings watching television in my dorm's common room. Often, as I lounged in front of the TV, a nagging thought tugged at my mind: "Why are you wasting time in front of the TV? Why don't you go upstairs and read your Bible?" I ignored this thought for quite a while, but one evening I finally gave in, turned off the set, went up to my room, and opened my Bible, which had begun to collect dust.

I knew better than to start in the Old Testament. I'd tried that before and quickly lost interest—a fact that I now find strange because I've come to love Genesis, not just for the history and science, but especially for its captivating stories and characters. This time, I started in the Gospel of John and was pleasantly surprised when I understood what I was reading. In fact, the words and stories started to come alive.

When I visited home on weekends, I'd ask my father questions about what I was reading and what it meant in both theological and practical terms. Week after week he listened thoughtfully to my questions, careful not to say too much or speak too quickly (a surprising response because he usually had much to say). At one point he gave me a small booklet focused on 1 Corinthians 3:10–15, which speaks of how Christ will judge each person's life, in grace and in truth. Insights from this booklet, combined with what I was learning from Scripture reading and these discussions with my father, had a profound impact on me.

For the first time, I clearly saw myself as a lost soul who needed a Savior. I recognized that I did not and could not, in my own strength, live up to God's moral standard, not even when I knew for sure what was right and wrong. Today, as I look back on my PhD research, my radio astronomy work, my publication and project successes, no discovery has been as critical as this recognition of my own moral failures. That's when I confessed my sin and asked God for his forgiveness. He gave it, and I received it, but I also realized that the Lord wanted me to make certain things right.

I made some confessions and apologies fairly quickly, but the stolen Boeing parts took some time (though I had never even used them in that hi-fi system!). I kept thinking, "When I visit my brother in Seattle, I'll go to Boeing and make it right." But I always found an excuse for not doing it. Then I would tell myself, "When I get home, I'll deal with it."

In the meantime, I finished my doctoral research on neutral hydrogen gas in galaxies beyond the Milky Way, and Caltech invited me to stay on for postdoctoral research in radio astronomy. During those years I married my wonderful wife, Diane, a nursing student I'd met at a Christian gathering. We found a place to live near the campus and began our family, first a son and then a daughter. When the opportunity arose for me to apply my radio astronomy research in the Netherlands, we decided to accept it. For two years we lived near the Westerbork Radio Observatory, where I helped set up a new radio astronomy array. While in Holland, we added another son to our family, and the five of us enjoyed many weekend adventures in Europe. Upon completion of my work at Westerbork, we moved back to California and back to Caltech, where I accepted a senior research fellowship in radio astronomy.

Clearing My Conscience

Back in the Pasadena area for good, Diane and I had a third son. I also resumed my involvement in a local church, where I taught a Bible study class for adults. As I was preparing a lesson one evening, God reminded me of my unfinished business with Boeing. How could I teach others about the importance of keeping a clear conscience when I didn't have one myself? So, before I could do anything else—eat, sleep, or prepare the lesson—I wrote a letter, enclosed a check, and sent it to the personnel department at Boeing. That Sunday, I confessed to the class what I had done, both my theft and my long-delayed attempt to make amends, and I asked the class members to pray for me, and with me, about whatever response might come.

Several weeks later, I received a reply from Boeing. The writer said that no one in the corporate office could figure out what to do with my letter. However, because she (the writer) was known to be a Christian, the personnel manager had simply handed the letter to her. This letter then opened a wonderful door for her, she said. It gave her an opportunity to discuss her Christian faith with some of her non-Christian coworkers. What a thrill for me and for my class to see how God could use even a failure—if confessed and made right—for his good purpose!

Called to Help Others Succeed

After completing my research fellowship, I was hired by the Jet Propulsion Laboratory (JPL), where I could apply my radio astronomy knowledge and skills in significant ways not only for the advance of knowledge but also for the benefit of my country. No job on Earth could have been a better fit for me. I contributed to many fascinating projects, including development of Very Long Baseline Interferometry (VLBI) for spacecraft navigation and construction of a parallel supercomputer called Hypercube. Along the way, I was asked to serve as a group supervisor.

In embracing this leadership role, I discovered what truly means the most to me: helping others grow and succeed in their work. My aim became clear. I wanted to lead teams with humility, informed by my Christian faith, in hopes of influencing teammates to recognize their need for God and his goodness. The Old Testament character Abraham intrigued and inspired me. God blessed Abraham so abundantly that his neighbors couldn't help but notice (Genesis 21:22–34). They, too, were blessed by Abraham's presence among them, even when they didn't understand why. I prayed that my life would have a similar effect on my coworkers, and I believe God answered those prayers. One day, one of the managers on the supercomputer project commended my obvious concern for my teammates' personal lives. He said his nickname for me was "father confessor" because I was quick to confess and seek forgiveness after speaking or acting impatiently toward my teammates, something I tried to practice with my family members and friends, as well.

Saving the Galileo Mission

One especially significant project my team was called upon to help with was the Galileo mission in 1989, which encountered a potentially disastrous problem. NASA had launched a probe toward Jupiter intended to explore the gas giant's expansive system of moons. However, the whole project was jeopardized when the probe's primary antenna refused to open. NASA was struggling to communicate with the spacecraft. The probe was slated to arrive at Jupiter in two years. If we weren't ready with a solution to the antenna problem, Galileo would be a failed mission, wasting billions of dollars of research funding.

NASA responded by assembling a special team, called the "tiger team," tasked with devising a remedy. My group proposed using an array of ground-based antennas using VLBI techniques to enhance communication with the probe's alternate antenna. The VLBI radio astronomy technique uses two antennas located miles apart to measure spacecraft positions and aid with navigation. VLBI, a project I had helped develop in the early days of my JPL career, was at this time being deployed in NASA's Deep Space Network to track its spacecraft. By employing this system, along with modifications to the communications system, we hoped to raise the communication rate to a level sufficient for collecting all the data NASA had originally aimed to gather. This mission, the Galileo S-Band Mission, was an endeavor that required multiple teams if we had any hope of achieving functionality within the brief two-year timeframe.

Leading this effort represented a monumental challenge. While I had served as group leader on a variety of previous projects, this one would be much more demanding. I had to examine myself, carefully considering what qualities I would need to head up a project such as this one and how I could best honor God through my efforts. Jesus said that one must become a servant in order to be great (Matthew 20:26). So, I took this teaching to heart, asking myself whether I was willing to roll up my sleeves and do the toughest work, if needed, rather than just handing it off to those under me. I realized I'd have to lead by example, refrain from complaining, and accept no special treatment if I were to create and maintain a sense of cohesiveness within this already hard-working group.

Toward this goal, I treated the team to a meal at their favorite restaurant each time we reached one of our milestones. God reminded me to affirm examples of diligence when I saw them. I also let the team know I was praying for them, asking God to guide them individually and as a group. Some outright chuckled at this comment, but, in the end, they seemed to appreciate the support. In fact, the group had so much success in meeting one milestone after another that one team member mused, "It almost makes me think there is a God who's helping us."

Finally, when the two-year moment of truth rolled around and the Galileo probe arrived at Jupiter, my team's efforts, along with many others' hard work, proved a success. The communication worked, the mission was saved, and the probe went on to make some groundbreaking discoveries about Jupiter's atmosphere and moons.

RTB and Science-Faith Harmony

After the Galileo project, I decided to retire from JPL in order to devote more time to ministry. I wanted to focus the remaining years of my life toward advancing people's faith in Christ. In 2001, I left my position at JPL and went to work with a friend from my Caltech days, Hugh Ross, and his ministry, Reasons to Believe (RTB). My team at JPL urged me to stay, but they understood how much my faith meant to me. At RTB, I took a much-needed administrative role as executive vice president and committed myself to building teamwork, using skills I had honed at JPL.

RTB's science apologetics work intrigued me. To be honest, I had never considered the dynamics between science and my Christian faith to be in conflict. I never really gave it much thought until I met Hugh Ross, my colleague in the 1970s. Hugh had grown up among secularists who, for the most part, saw science, not the Bible (or any religious belief system), as the only reliable source of answers to life's big questions. Unlike anyone I had ever met, Hugh made a thorough search for the answers—and found them where he least expected them, in a synthesis of scientific fact and biblical revelation.

In a close and careful reading of the Old and New Testaments, Hugh saw a harmony and consistency with scientific findings, including some of the latest discoveries in his field of astronomy. Within the context of my faith, I had always taken for granted that science and physics offer countless indicators of a Creator's necessity: the beginning of the universe, the fine-tuning observable in the universe, the overwhelming evidence for design, etc. Over the years, I had engaged in conversation with many people from various backgrounds who pointed to ideas such as the multiverse as a basis for rejecting belief in a transcendent Creator. I found Hugh's books, beginning with *The Fingerprint of God*, helpful in showing skeptics a scientifically credible basis for the Christian faith and worldview. It has been my joy to work with him and with the team that formed around him to this day, even long after handing off my full-time role and responsibilities to others.

The Most Important Thing

However, when it comes to what for me constituted the most convincing case for Christianity, I always come back to that moment when I saw myself as a sinner in need of a Savior. The Scriptures exposed who I really am, and it's dead accurate in its description. Although there is much to learn about the Bible—how it was written and its accuracy preserved through the ages, etc.—in the final analysis, it's a book that calls me to be just and good, and I simply cannot be that on my own. That's why I needed—and still need—Jesus Christ. In the words of my favorite author, C. S. Lewis,

> The Christian is in a different position from other people who are trying to be good. They hope, by being good, to please God if there is one; or—if they think there is not—at least they hope to deserve approval from good men. But the Christian thinks any good he does comes from the Christ-life inside him. He does not think God will love us because we are good, but that God will make us good because He loves us; just as the roof of a greenhouse does not attract the sun because it is bright, but becomes bright because the sun shines on it.[1]

I am convinced beyond all measure that the most important thing in life is to have God's blessing. "For the eyes of the LORD range throughout the earth to strengthen those whose hearts are fully committed to him" (2 Chronicles 16:9). There is nothing that matters so much to me as to have this support, this "blessing," from God.

In the end, I trust that by God's great mercy and grace, my life will have served his kingdom well. I want to leave a legacy that brings glory to him who is my Lord and my Savior.

Chapter 12

God's "Two Books" Still Speak

Hugh Ross

When asked to relate how I became a follower of Jesus Christ, I often begin by saying, "Science led me to him." This simple statement has surprised countless audience members over my decades of public presentations because many people have thought the opposite was the case. They've told me that science should lead any thinking person *away* from the mythical God of the Bible and *toward* a factual view of reality. From my earliest studies of both the record of nature and the words of the Bible, I didn't see this apparent science-faith conflict. I contend that it's been perpetuated needlessly. Everyone's life experience is different, but I hope that by sharing some of my background, you'll see that God speaks truth to human beings in both of his books: nature and Scripture.

I didn't get to know Christians until many years after I became one. That's not to say that during my pre-Christian years God did not work through people, as well as through circumstances he clearly orchestrated for me. My hope in offering this detail is to demonstrate how significant and effective our prayers and personal interactions can be in the life of people we may never get to know.

When I say I didn't really know Christians, I mean I did not know any well enough to engage in meaningful spiritual conversations with them. That experience eluded me until about eight years *after* I committed my life to Christ. However, several individuals did, indeed, play critical roles in my journey to faith—including non-Christians who got my attention and spoke to me along the way.

Early Family Hardship
I was born in Montreal, forty days before Japan's surrender brought World War II to an end. My father, a self-taught engineer from Calgary, served the war effort by designing and manufacturing hydraulics for Allied Aircraft. At war's end, he became one of two partners in a rapidly growing hydraulics engineering firm. However, my father's partner became enticed by the company's quickly accumulated wealth and he seized the liquid assets and fled the country, leaving my father with a bankrupt enterprise. My father responded by draining his own personal savings to provide each employee with a final paycheck.

Having completed his formal education only up to the tenth-grade level, and now facing a ruined business reputation, my father saw no realistic hope of recovery, especially for the support of his young family, in the Montreal area. So, he used his last few hundred dollars to move my mother, my two sisters (ages one and two) and me (age four) to western Canada.

My parents' circle of friends in Montreal, including some of my mom's nurse colleagues, had observed my seriously lagging development of social, motor, and language skills. They became convinced I was "mentally retarded" (their term, not mine) and encouraged my parents to place me in an institution. Before my parents could give this recommendation any serious consideration, they found themselves in an immigrant neighborhood in Vancouver. Autism spectrum disorders were as yet unrecognized, certainly unknown to my parents and the healthcare community. I can easily see how my withdrawn behavior, unexpressive face, limited verbal communication, and lack of typical motor development led my parents' friends in Montreal to draw their conclusion.

One Special Teacher
In the years right after World War II, the huge influx of refugees from Europe and Asia led to an extreme housing shortage in Canada, but my parents were determined to find a house for our family. Finding no other options available, they bought a condemned house for $6,000 in what was then one of Vancouver's poorest neighborhoods. We moved into it with just six weeks remaining of my first-grade school year.

I was panic-stricken. I was failing in all my subjects. I was unable to control a pencil to make legible letters and numbers. Because I had made no effort to speak, my communication skills were seriously underdeveloped. How could I prove to anyone that I could read and do arithmetic?

My new first grade teacher, Lila Campbell, somehow discerned the fear and frustration in my eyes. With just a few days left in the school year, she kept me after class one day and waited for all the other children to leave. She said, "I am going to ask you questions about these books from the shelf in our classroom. You don't need to talk. Just nod your head for *yes* and shake your head for *no*." By this means, she determined I had read and understood several of the books. She had no way of knowing I was reading *David Copperfield*, a tattered copy of which I found stuffed inside the wall of our still derelict house.

On the last day of school, Miss Campbell read the names of our class members who would pass into second grade. I focused on my shoelaces. My name was the second to last one she read. I knew from conversations I'd overheard between her and the principal that she'd gone out on a limb for me. I couldn't let her down. During that year's summer vacation, I spent hours each day practicing with my pencil, working to make legible letters and numbers.

On my first day of second grade, the teacher arranged for us to sit, as far as I could tell, in order of our first grade academic achievement. I occupied the last chair, and my classmates lost no time labeling me the class dummy. It didn't help that I still wasn't talking. That's when I resolved that no matter how badly I embarrassed myself, I would take whatever opportunities I was given to practice speaking.

After every set of tests, our teacher reseated us, and with each reseating, I found myself several seats closer to the front of the class. By the end of second grade I sat in the first chair.

When I was 34, during a visit to my parents in Vancouver, I received a surprising invitation. Miss Campbell, who had never lost track of my family, invited me and my wife to her home for tea. During that afternoon, Miss Campbell revealed that I was her "mystery" student, the one who most puzzled her. She had followed my progression from the bottom of the class to the top in second grade. In fact, from second grade onward she had kept track of my academic progress. What's more, she revealed she had been praying for me since our first meeting back in her first grade classroom. Yes, she was a Christian!

Miss Campbell said she had prayed I would use my education and intellect in service to Christ. Kathy and I had the opportunity to thank her profusely for the remarkable ways (both known and, until that moment, unknown) God had used her to shape my destiny. I cannot imagine my life's trajectory without the intervention and intercession of this dear Christian lady I had been in contact with for just six weeks.

Introduction to Libraries
One of the privileges of graduating into second grade was access to the school's

library. Within the first few weeks I read dozens of books on history, geography, and science. One was Fred Hoyle's newly published *The Nature of the Universe*, wherein I read, "There is a good deal of cosmology in the Bible."[1] Hoyle described it as "a remarkable conception."[2] This comment piqued my curiosity, but throughout the rest of the book he expressed a thoroughly negative opinion of the Bible and Christianity. Nevertheless, I did not forget his words about cosmology in the Bible.

My second grade teacher fortuitously organized a class field trip to the Vancouver Public Library, which at that time housed nearly three million books. She helped us find the bus routes and transfers to get to the library on our own, as long as our parents allowed. She also saw to it that each of us received a card to the library's substantial children's collection.

After that field trip, I spent nearly every Saturday at the Vancouver Public Library, checking out the maximum-allowed five books, almost always on physics and astronomy. Having exhausted the physics and astronomy books in the children's section, I somehow managed to persuade one of the librarians to trust me with a card for the adult section. By age eight, I had decided my future career would lie in some discipline of astronomy. Eventually, I was granted access to the library at the University of British Columbia (UBC).

Scientific Method

In the public schools I attended, teachers taught, retaught, and reviewed the scientific method every year from first grade through twelfth. I had been trained to apply the scientific method not only to every science question I encountered but also to any subject matter open to analysis or interpretation. Not until many years later, though, did I learn that the scientific method originated and took shape among devout clergymen and theologians who devoted themselves to close and careful study of both the Bible and the natural world.

Gideon Bible

When I was in the sixth grade, two businessmen paid a visit to my public school. They placed two boxes on our teacher's desk, smiled, and left without saying a word. The teacher informed us there were Bibles in the boxes, and any student who wanted one could come and take one. I never passed up the chance to get a free book.

I put my Gideon Bible on the bookshelf in my room, and that's where it stayed, strangely unopened, for the next six years. However, during those six years, thanks to my English teachers' love of William Shakespeare, I became *almost* fluent in—or at least unintimidated by—the language of King James. So, at age 17, when I had become convinced by my studies in physics and astronomy that a Creator-God must exist, I was able to read and understand the King James Version of the Bible I had received.

Cosmic Beginnings

My reading in cosmology had persuaded me, by this time, that the universe had a beginning. A cosmic beginning implies, by no great stretch of logic, the necessity of some sort of cause (or beginner). My curiosity drove me to learn what I could about the cosmic beginner. Given that astronomers frequently referred to Immanuel Kant as the father of cosmology, I started with Kant's books, especially his *Critique of Pure Reason*.[3]

I found that much of the book's content was devoted to discounting any and all cosmological evidence for God. Kant's approach called for chopping up the cosmological evidence into tiny pieces and subsequently demonstrating that each piece, by itself, offered insufficient proof for God. This line of reasoning seemed disingenuous, or at least agenda-driven. Hence, I didn't see any reason for doubting the harmony between science and faith. What's more, Kant's expressed assertions about space and time and the principle of cause and effect[4] were counter to well-established science and the newly emerging space-time theorems.[5]

Putting aside Kant, I picked up the works of other highly regarded philosophers. None brought me any closer to an explanation of the realities observed and measured through increasingly powerful instruments—realities of a transcendent cosmic origin and of ongoing fine-tuning required to make a planet like Earth and its panoply of life-forms possible. I found these books just as disappointing as Kant's. They seemed to raise more questions than they answered—questions other than those I was asking.

Testing the World's Holy Books

At this point, I was nearly convinced I would be compelled to live with the mysterious unknown—perhaps unknowable—beginner. For diversion, I read a book of creation stories from many different cultures. They seemed colorful and intriguing but, as expected, more fantastic than factual. Nevertheless, I felt that to be thorough and fair-minded, I should investigate the "holy books" and commentaries of the world's great religions. Perhaps each one held some element of truth that, when combined, would present a somewhat coherent picture.

Picking up one after another, I found some wise words about one topic or another, but I also encountered vague, esoteric language and violations of confirmed fact. In some cases, I encountered outright rejection of material reality. My greatest disappointment came from finding so much repetition and so little material that could be tested for veracity.

Testing the Bible

Finally, more than halfway through my seventeenth year, I picked up the Gideon Bible that had been collecting dust on my bookshelf for the past six years. Again, thanks to my teachers' love of Shakespeare, I had no trouble understanding its archaic English. To my astonishment, the scientific method leapt out from the very first page. There I saw a clear statement of the text's frame of reference and of Earth's initial conditions (Genesis 1:2) [now established], followed by an accurate chronological overview of Earth's and life's development (Genesis 1:3–27).[6] The text also provided an answer to the fossil record enigma that had bothered me for some time.

This enigma—the appearance of abundant new phyla, classes, orders, and families of life before the arrival of humanity and not after—can be explained if, in fact, the Creator has a purpose and plan for humanity. I noted that for six prehuman creation periods, God apparently intervened to prepare the environment and introduce new life-forms, all for the ultimate benefit of human life. Then, during the seventh "day" (the human era), God ceased from creating new kinds of life to focus on the one creature formed uniquely for relationship with a transcendent Being. The mystery of life's early origin on Earth and of the rapid emergence of increasingly complex life from the Cambrian explosion right up to the appearance of humanity found resolution in the possibility of purposeful intervention. And that possibility is rooted in the reality of a transcendent cosmic origin, which my astrophysical research had been pointing toward for some time.

Not one to be credulous, I spent the next 18 months testing as many of the Bible's statements relevant to science, history, and geography as I could find. The text gave me much content to test and even encouraged testing (1 Thessalonians 5:21)—a fact that, in itself, kept me going. Night after night I studied in the privacy of my room after finishing problem sets for my physics classes. I searched for demonstrable errors or contradictions and found none. I did find paradoxes and passages that I didn't fully understand, but nothing that I could demonstrate with certainty as a contradiction or error. Instead, I found dozens of passages where the Bible accurately predicted (meaning it described what we would expect to observe if a particular idea were true) future scientific discoveries and historical events, some even thousands of years ahead of their time.

For example, I noticed that Genesis 1 matched both the chronological order and the scientific description of major events in Earth's history (Genesis 1:1–27). Statements in Ecclesiastes, Jeremiah, and Romans described the unchanging nature of the laws of physics throughout the universe (Ecclesiastes 1, 10–12, Jeremiah 33:25, Romans 8:20–22). Daniel's prophecies of the rise and fall of future empires (Daniel 2:26–45, 7:1–8:27, 10:20–11:45) and Ezekiel's prophecies concerning the rebirth of Israel as a nation proved historically accurate (Ezekiel 33:27–37:28). Such amazing and consistent predictive power, I deduced, could be explained only if the Bible authors, who spanned a considerable breadth of time, location, and culture, had been inspired by the One who brought the universe into existence.

Not a Humanly Crafted Book
I also noted that the Bible, unlike other books held sacred by various cultures, described attributes of God and other truths that defied explanation within the limited length, width, height, and time dimensions we humans experience and yet could hold true in an expanded dimensional frame. These attributes and doctrines require the existence of dimensions (or their equivalent) beyond the ones we experience, or a Being who exists and operates beyond the confines of dimensions. Given that humans can visualize phenomena only in the dimensions they experience (despite mathematicians' ability to play with extra dimensions), the biblical references to these attributes and doctrines suggested to me that the Bible, unlike any other book, could not have sprung from mere human imagination.

This accumulation of observations and conclusions—that the Bible was free of historical and scientific error, accurately predicted future events and discoveries, and described attributes of God that superseded human visualization capacity—seemed simultaneously thrilling and chilling. Although I still had unanswered questions, I could not ignore the personal implications of what I had discovered.

Commitment to Christ
What if God's plans for me differed from the ones I had laid for myself? How would my parents and sisters, professors and friends react to my new belief that I owed my life, my entire being, and future to God in gratitude for what Jesus

Christ, the eternal Word, had done for me through his sinless life, atoning sacrifice, and bodily resurrection? My life would no longer be my own to do with as I pleased. The Gideons had summarized all this and more for me in the back of that little Bible. They also provided a place for me to sign and date, sealing the transfer of authority over my life.

Fear and pride held me back for a while. The anticipation of ridicule and rejection loomed large, not to mention awareness of my own weaknesses and deficits. For a few weeks I tried living like a Christian without actually *being* a Christian, as if it can be done. What a hollow experience that was! The more I tried to keep my thoughts and attitudes pure, the more glaringly I failed. Even my academic performance took a brief downturn, something I had not experienced since the start of second grade. Suddenly I recalled the words at the beginning of the book of Romans, words describing how rejection of truth leads to a darkening of the mind (Romans 1:21).

The moment of decision had come. I sat on my bed, perspiring, and asking God to make me humble enough to surrender to him, but my tension only increased. Finally, it occurred to me that humbling myself was my part, and God would take me from there. An enormous wave of relief and peace rolled through me as I signed my name and recorded the time that I unreservedly committed my life to Jesus Christ: August 7, 1964; 1:05 AM.

First Encounter with a Skeptic

The timing of my decision to entrust my life to Christ now seems more strategic than accidental. My physics lab partner throughout my sophomore, junior, and senior years at UBC was an astute thinker and careful researcher named John Samson. John went on to become a geophysicist and for several years served as chair of the physics department at the University of Alberta. The night I signed my name in the back of my Gideon Bible, John came to mind. I knew that my commitment to being Christ's disciple meant telling others what I had done and inviting them to do the same. Like me, they would need assurance that Jesus Christ is, indeed, the Creator of everything, that the Bible is true, and that the Good News the Bible proclaims about God's purpose and plan to redeem us can be trusted.

John was the first person I wanted to talk with about my newfound faith in Christ. Yet with him my intimidation factor loomed high. I still struggled somewhat with verbal fluency, and John made a habit of challenging everything I said until I could demonstrate to his satisfaction that it was true. With trepidation overcome by a desire to let my friend in on a life-giving discovery, one afternoon after our last physics class of the day I tried to get a spiritual conversation started.

Before I could even put one sentence together, John stopped me. He said, "Hugh, I can see you want to talk about something, but I really need to talk. I need to talk to someone about God. Do you know of anyone on this campus who knows anything about God?" For the next two hours John and I talked about God, the Bible, the Christian faith, and my very recent commitment to becoming a follower of Christ. John questioned me closely about how I became convinced that the Bible was supernaturally inspired and totally trustworthy. John was especially intent on testing the Bible's predictive power. When I told him that, for example, the second rebirth of Israel as a nation along with other developments in the modern era precisely fulfilled biblical prophecy, John, true to form, said, "Prove it."

Two evenings later, we were both working in the main university library on three sets of exceptionally difficult physics problems. Strangely, an assignment we thought would take until two or three in the morning to complete took each of us only till midnight. Given this unexpected availability of time, I asked John if he'd like to see some of the evidence he had asked for.

We went to the part of the library where microfilm archives of major newspapers were stored. We spent the next two hours comparing biblical prophecies about Israel with newspaper stories about developments in Israel published between 1945 and 1960 in the *Jerusalem Post*, *San Francisco Chronicle*, and *The Times* (London).

John saw with amazement how the newspaper reports aligned with the Bible's prophetic statements. We spent other late-night sessions scanning the library's microfilm and talking about God and physics. Although John never made clear to me whether he embraced God's offer of redemption in Christ, I did observe that he no longer mocked Christians or Christianity. Years later, I met Christians who had been his students at the University of Alberta. They told me John had encouraged their faith.

Divine Appointments

My encounter with John drove home to me the reliability of the promise in Ephesians 2:10, which says God has prepared, ahead of time, good things for us to do. It also underscored God's admonition in 1 Peter 3:15–16 to always be ready to give reasons for our hope in Christ. Initially, I thought I had to be the one to orchestrate a spiritually significant encounter, but that day I saw that God had already arranged one for me. I thought I would have to find a way to steer the conversation toward my reasons for hope in Jesus Christ, but as soon as I became prepared to offer reasons "with gentleness and respect," God began to lead me to people who would ask me for those reasons.

Each year my joy in knowing Jesus Christ and in sharing the Good News of what he has done and will do for those who look to him for Life with a capital "L" grows greater. Each year, the growing wealth of scientific data adds to the immense body of evidence for God's existence and purposes for creating human beings for relationship with him. Each year brings new reasons to believe in him and in his plan for the future, the promised creation yet to come.

Science-Faith Harmony Grows

Some people I meet express apprehension about connecting the constantly emerging discoveries about our natural world—a perpetually open book—with the "closed" book of Scripture. Two responses come immediately to mind. First, having read the Bible over and over since I first opened its pages at age 17, experience tells me that neither I nor any other person can ever fully plumb its depths. Something fresh emerges from every reading. It's as though "the Bible reads me," as others have so aptly commented.

Second, in my decades of continuing to survey the latest scientific literature, I've found that with each new advance, each new breakthrough in our understanding of the cosmos, evidence for the reliability of Scripture and, thus, for my faith in Christ increases. The alignment between God's "two books," nature and the Bible, remains as strong as ever or grows even stronger. For example, fossil evidence initially suggested that vegetation on the landmasses followed, rather than preceded, animal life in the seas, in contradiction to the Genesis creation chronology. However, ongoing research over the past decade brought confirmation that vegetation on the landmasses did, indeed, come first.[7]

Meanwhile, countless media headlines continue to proclaim the demise of the big bang. The notion that all matter, energy, space, and time sprang into existence roughly 14 billion years ago seems disquieting to those who recognize the philosophical implications. And yet, rather than overturning our understanding of the cosmic creation event, new revelations from continually advancing technology—from the WMAP and Planck satellites to the amazing James Webb Space Telescope—further affirm or refine our big bang creation models.

For my writing and speaking, I find that I must choose which discovery to highlight from a host of exciting discoveries. Evidence mounts continually for the guidance of a purposeful power in shaping a unique home for humanity.

For example, a series of findings recently showed how our unique planet-moon system with a strong, coupled magnetosphere for the first half billion years of its formation prevented our star, the Sun, from sputtering away all of Earth's atmosphere and surface water.[8] Another example comes from the discovery that viruses, which some say counter the Genesis claim of a "good" creation, play a crucial role in enhancing and regulating Earth's water, carbon, and oxygen cycles so that advanced life is possible.[9] Even our galaxy's supermassive black hole speaks of divine provision and protection. It is 35 times smaller than the black hole at the core of other spiral galaxies of similar mass, and during the human era it has been exceptionally "quiet" in terms of the quantity of radiation emitted from beyond its event horizon.[10] Mounting discoveries add further confirmation of the complete harmony and constructive integration of God's two books.

The thrill of discovery never ends for me. Never does research fail to provide new mysteries to explore and a greater sense of awe for the one who speaks in both actions and words. The more we learn about nature—on whatever size scale, inorganic to organic—the more evidence we uncover for the supernatural handiwork of the biblical Creator and Redeemer and the author of both books.

Chapter 13

Conclusion: An Invitation to the Journey

Krista Bontrager and George R. Haraksin II

The stories presented in this book are written by members of Reasons to Believe's (RTB's) Scholar Community. This unique network of Christians works professionally in various STEMM fields and in other disciplines such as theology, philosophy, and law. They share a common love for exploring the creation and knowing its Creator. This book is a compilation of their journeys to understand the creative interaction of science and faith and wrestle with the so-called "war" metaphor or conflict thesis that so often characterizes its relationship.

One of our hopes in assembling these stories is that they'll inspire you to consider beginning your own journey to explore the intersection of science and the Christian faith. We imagine that this might seem like a daunting task. But by reading this book, you're already off to a good start. The authors have laid a solid foundation for you to continue the next leg of your own journey.

Maybe you've never considered God. Maybe you aren't even looking for him. Maybe you've just been so busy in your research pursuits that you haven't had time to investigate whether God is real or whether the Bible's claims might be true. Maybe you're mad at God because you feel like he didn't show up for you at a key moment of suffering in your life.

Hugh Ross shared about his decision to read the Bible and compare its claims about the physical world with the findings of modern science. We invite you to consider doing the same: to pick up a Bible and read Genesis 1–3 and the Gospel of Matthew or John. As you read, you can record observations and questions that you want to investigate. After that, focus on two or three specific issues that you want to pursue in a deeper way. You can use the search bar on our website at reasons.org as part of your research. We may have an article or podcast related to your questions.

All of this may sound challenging, but keep in mind that just as you've probably spent years becoming an expert on a particular issue, it may also take time to get oriented to investigating an entire religion. And you don't need to figure everything out all at once.

Perhaps you're already a Christian, but this book has provoked your thinking to consider how your area of expertise may connect to your faith. We invite you to reach out to RTB and explore the relationship between science and the Christian faith.

This process can also involve its own set of challenges. Perhaps you've been taught that the Bible teaches that the universe is only 6,000–10,000 years old. If that's the case, then you've probably been wondering how the authors of this book have made peace between science and faith. *Are they just downplaying certain details from the Bible?* Our approach to integrating the words of the Bible with the record of nature rests on robust investigation. We look for points of agreement between these two data sets. And where they appear to disagree, we dig deeper into the details. We are confident that the Creator of both the Bible and nature (the author of these "two books") will reveal himself consistently, and we proceed from that starting point. The start of your journey into investigating the harmony between the Bible and nature can be as simple as visiting our website. We offer an array of podcasts, articles, and courses to help you dig deeper.

If you have a terminal degree in science, theology, philosophy, or medicine, we invite you to consider whether joining RTB's Scholar Community might be a fit for you. We're building a network of hundreds of Christian professionals working in STEMM fields and beyond who want to practice sound science with a testable faith. They seek to reveal God through and in science wherever they find themselves. If you'd like to find out more about this community, visit the RTB website and look for the Scholar Community page.

Our hope at RTB is to bring a lasting peace to the "war" between science and faith. It's a war that never should have started in the first place and peace is real, if you want it.

Notes

Introduction: Rumors of War (and Peace)

1. Jerry Coyne, "Yes, There Is a War Between Science and Religion," The Conversation, December 21, 2018, theconversation.com/yes-there-is-a-war-between-science-and-religion-108002.
2. Following World War II's ending, Lieutenant Hiroo Onoda of the Japanese Imperial Army refused to surrender until March 1974, even after the Japanese government dropped explanatory leaflets over his hiding area. See Evan Andrews, "6 Soldiers Who Refused to Surrender," History Channel, published February 26, 2013, updated December 13, 2022, history.com/news/6-soldiers-who-refused-to-surrender.
3. John Lennon and Yoko Ono, "Happy Xmas (War Is Over). (Ultimate Mix, 2020) John & Yoko Plastic Ono Band + Harlem Community Choir," posted August 24, 2010, YouTube video, 3:33, youtube.com/watch?v=yN4Uu0OlmTg.
4. Cary Funk and David Masci, "5 Facts About the Interplay between Religion and Science," Pew Research Center, October 22, 2015, pewresearch.org/fact-tank/2015/10/22/5-facts-about-the-interplay-between-religion-and-science/.
5. John William Draper, *History of the Conflict Between Religion and Science* (D. Appleton and Co., 1878).
6. Andrew Dickson White, *A History of the Warfare of Science with Theology in Christendom*, 2 vols. (Macmillan, 1896).
7. Timothy Larsen, "'War Is Over, If You Want It': Beyond the Conflict between Faith and Science," *Perspectives on Science and the Christian Faith* 60, no. 3 (September 2008): 147–155, asa3.org/ASA/PSCF/2008/PSCF9-08Larsen.pdf.
8. David Kinnaman and Gabe Lyons, *unChristian: What a New Generation Really Thinks About Christianity . . . and Why It Matters* (Baker Books, 2007), 55–56.
9. David Hutchings, "Flat Wrong," in David Hutchings and James C. Ungureano, *Of Popes and Unicorns: Science, Christianity, and How the Conflict Thesis Fooled the World*, online ed. (Oxford Academic, 2021), accessed April 6, 2023, doi:10.1093/oso/9780190053093.003.0003.
10. Hutchings and Ungureanu, *Of Popes and Unicorns*.
11. Rudolf Otto, *The Idea of the Holy*, trans. John W. Harvey, 2nd ed. (Oxford University Press, 1950), 7–11.
12. Alan Lightman, *The Transcendent Brain: Spirituality in the Age of Science* (Pantheon, 2023), 3–5.
13. Alex Gomez-Marin, "Making Sense of the Sacred," *Science* 380, no. 6640 (April 6, 2023), doi:10.1126/science.adh0532.
14. Lightman, *The Transcendent Brain*, 8–9.

Chapter 1: A Match Made in Heaven

1. In physics, the principle of relativity is the requirement that the laws of physics (as codified in equations) must have the same forms in all admissible frames of reference.
2. Jeff Zweerink, *Who's Afraid of the Multiverse?* (Reasons to Believe, 2008).
3. For some examples, see Erik A. Petigura, Andrew W. Howard, and Geoffrey W. Marcy, "Prevalence of Earth-Size Planets Orbiting Sun-Like Stars," *Proceedings of the National Academy of Sciences* 110, no. 48 (November 4, 2013): 19273–19278, doi:10.1073/pnas.1319909110; Steve Bryson et al., "The Occurrence of Rocky Habitable-Zone Planets Around Solar-like Stars from Kepler Data," *Astronomical Journal* 161, no. 1 (December 22, 2020): id. 36, doi:10.3847/1538-3881/abc418; and Michelle Kunimoto and Jaymie M. Matthews, "Searching the Entirety of Kepler Data. II. Occurrence Rate Estimates for FGK Stars," *Astronomical Journal* 159, no. 6 (May 4, 2020): id. 6, doi:10.3847/1538-3881/ab88b0.
4. For specific quotes from Galileo and Kepler, see Michael J. Crow, ed., *The Extraterrestrial Life Debate: Antiquity to 1915* (University of Notre Dame Press, 2008), 52, 60–61. For more on the theological significance of finding intelligent ET, see Ted Peters, "The Implications of the Discovery of Extra-terrestrial Life for Religion," *Philosophical Transactions of the Royal Society A*, 369 (January 10, 2011): 644–655, doi:10.1098/rsta.2010.0234.

Chapter 2: Finding Answers in Christ

1. Andrew Dickson White, *A History of the Warfare of Science with Theology in Christendom*, 2 vols. (Cosimo Classics, 2009).
2. Nancy R. Pearcey and Charles B. Thaxton, *The Soul of Science: Christian Faith and Natural Philosophy* (Crossway, 1994).
3. Norman Geisler and Peter Bocchino, *Unshakable Foundations: Contemporary Answers to Crucial Questions About the Christian Faith* (Bethany House, 2001), 71–74, 113–114.
4. J. P. Moreland, *Scaling the Secular City: A Defense of Christianity* (Baker Academic, 1987), 211.
5. Psalm 19 helps us discover God's fingerprints in creation and his specific will for us through his Holy Word.
6. Francis A. Schaeffer, *The Complete Works of Francis A. Schaeffer*, vol. 2, *A Christian View of the Bible as Truth* (Crossway, 1985), 139.
7. C. Stephen Evans, *Pocket Dictionary of Apologetics and Philosophy of Religion* (InterVarsity Press, 2002), 114.
8. Fazale R. Rana with Kenneth R. Samples, *Humans 2.0: Scientific, Philosophical, and Theological Perspectives on Transhumanism* (RTB Press, 2019), 20.
9. Paul Brand and Philip Yancey, *The Gift of Pain: Why We Hurt and What We Can Do About It* (OM Books, 1999), 192.
10. Brand and Yancey, *Gift of Pain*, 3.
11. Brand and Yancey, 5.
12. Brand and Yancey, 187.
13. C. S. Lewis, *Mere Christianity* (Macmillan, 1960), 76.
14. Charles W. Colson, *Justice That Restores: Why Our Justice System Doesn't Work and the Only Method of True Reform* (OM Books, 2001), 23.
15. Gary R. Habermas and Michael R. Licona, *The Case for the Resurrection of Jesus* (Kregel Publications, 2004), 52.
16. Gary R. Habermas, "The Historical Jesus" (lecture for the Christian Apologetics Program, Biola University) audio CD, disc 1, 63:08 (Biola University Christian Apologetics Program,

January 1, 2005).

17. Michael Green, *The Day Death Died* (InterVarsity Press, 1982), 14.

Chapter 3: Testing Science, Faith, and Spiritual Experience

1. Blaise Pascal, "Memorial" (Christian Classics Ethereal Library), accessed March 18, 2023, ccel.org/ccel/p/pascal/memorial/cache/memorial.pdf.
2. Blaise Pascal, *Pensées* (Penguin Books, 1995).
3. Pascal, "Memorial."
4. Acts 9:1–19.
5. Pascal, *Pensées*, 121–125.
6. Gerald Schroeder, *The Science of God: The Convergence of Scientific and Biblical Wisdom* (Broadway Books, 1997).
7. Schroeder, *The Science of God*, 41–59.

Chapter 4: An Obstetrician's Reflections on Life and Suffering

1. Women living in underdeveloped countries without access to a caesarean section can labor for days without progress. The trauma of prolonged obstructed labor can result in an abnormal connection or fistula between the bladder and the vagina. Instead of having control of their bladder, these women leak urine continuously through the vagina. Because of this, their husbands often desert them.
2. Gary C. Schoenwolf et al., *Larsen's Human Embryology*, 5th ed. (Elsevier Saunders, 2015), 51.
3. R. L. Bree et al., "Transvaginal Sonography in the Evaluation of Normal Early Pregnancy: Correlation with HCG Level," *American Journal of Roentgenology* 153, no. 1 (August 1989): 78, doi:10.2214/ajr.153.1.75; Peter M. Doubilet et al., "Diagnostic Criteria for Nonviable Pregnancy in the First Trimester," *New England Journal of Medicine* 369, no. 15 (October 10, 2013): 1448, doi:10.1056/NEJMra1302417.
4. This paragraph uses embryologic gestation age, which is the age from conception. Most physicians use menstrual age, the age from the last menstrual period or about two weeks prior to conception. I have routinely seen cardiac activity at six weeks of gestation (four weeks of embryologic age).
5. Nancy R. Pearcey, *Love Thy Body: Answering Hard Questions About Life and Sexuality* (Baker Books, 2018), 25.
6. Eszter Fanczal et al., "The Prognosis of Preterm Infants Born at the Threshold of Viability: Fog over the Gray Zone—Population-Based Studies of Extremely Preterm Infants," *Medical Science Monitor* 26 (December 10, 2020): e926947-1-12, doi:10.12659/MSM.926947; Matthew A. Rysavey et al., "Between-Hospital Variation in Treatment and Outcomes in Extremely Preterm Infants," *New England Journal of Medicine* 372, no.19 (May 7, 2015): 1806–1808, doi:10.1056/NEJMoa1410689.
7. Pearcey, *Love Thy Body*, 25.
8. K. B. Kuchenbaecker et al., "Risks of Breast, Ovarian, and Contralateral Breast Cancer for Brca1 and Brca2 Mutation Carriers," *JAMA* 317, no. 23 (June 20, 2017): 2402–2416, doi:10.1001/jama.2017.7112.
9. C. S. Lewis, *The Problem of Pain* (Harper One, 1996), 31.
10. Lewis, *The Problem of Pain*, 40–41.
11. The poem has been attributed to various authors such as Grant Colfax Tullar, Benjamin Malacia Franklin, and Florence M. Alt. It can be found here: "The Weaver," Goodreads, accessed June 5, 2023, goodreads.com/quotes/741391-the-weaver-my-life-is-but-a-weaving-

between-my.

Chapter 5: How Could a Jew Believe in Jesus?

1. D. L. Block, "General Relativity, and Its Applications to Selected Astrophysical and Cosmological Topics," *Quarterly Journal of the Royal Astronomical Society* 15 (September 1974): 264–291.
2. Cornelia Sussman and Irving Sussman, "Marc Chagall, Painter of the *Crucified*," in *The Bridge: A Yearbook of Judeo-Christian Studies*, ed. John M. Oesterreicher and Barry Ulanov, vol. 1 (Pantheon Books, 1955), 115.
3. Jakób Jocz, "Is It Nothing to You?," *Church Missions to Jews* (1941), 8, wycliffecollege.ca/jocz.
4. Yosef Haim Brenner, writing in the Zionist newspaper *Hapoel Hatzair*, November 24, 1910. Original Hebrew article available at benyehuda.org/read/867.
5. Sussman and Sussman, "Marc Chagall," 103.
6. Fyodor M. Dostoevsky, *The Grand Inquisitor*, trans. S. S. Koteliansky (E. Mathews and Marrot, 1930), 5–6.
7. By *Dixi*, he means, "I have said all that I have to say, and thus the argument is settled." Dostoevsky, *Grand Inquisitor*, 29.
8. A particularly joyous occasion was to fly into the Seychelles archipelago in 2014, the year I turned 60, for an international astronomy conference. Senior colleagues included Ken Freeman and Giovanni Fazio. For the conference proceedings, see Kenneth Freeman et al., eds., *Lessons from the Local Group: A Conference in Honour of David Block and Bruce Elmegreen* (Springer, 2015).
9. A *rhema* is a verse (or portion of Scripture) that the Holy Spirit brings to our attention. We may need revelation into a current situation we are facing, or we may need direction (in my case, direction into which field of research God wanted me to undertake). The Holy Spirit guides us (John 16:13).
10. David L. Block, "NGC6872: A Remarkable Barred Spiral Galaxy," *Astronomy and Astrophysics*, vol. 79 (Springer, 1979), L22–L23, ui.adsabs.harvard.edu/abs/1979A%26A....79L..22B/abstract.
11. The human backbone, which encloses the spinal cord and its surrounding fluid, cannot be seen optically. We need new eyes to penetrate our skin.
12. D. L. Block and R. J. Wainscoatt, "Morphological Differences Between Optical and Infrared Images of the Spiral Galaxy NGC309," *Nature* 353, no. 6339 (September 1991): 48–50, doi:10.1038/353048a0.
13. D. L. Block et al., "Imaging in the Optical and Near-Infrared Regimes II. Arcsecond Spatial Resolution of Widely Distributed Cold Dust in Spiral Galaxies," *Astronomy and Astrophysics* 288 (August 1994), 383–395.
14. David L. Block and Kenneth C. Freeman, *God and Galileo: What a 400-Year-Old Letter Teaches Us About Faith and Science* (Crossway, 2019).
15. In this sense, Blaise Pascal speaks of the "hidden God" (Latin: *Deus absconditus*).
16. John Paul II, "Allocution of the Holy Father John Paul II to the Participants in the Plenary Session of the Pontifical Academy of Sciences," October 31, 1992, in *Papal Addresses to the Pontifical Academy of Sciences 1917–2002, and to the Pontifical Academy of Social Sciences 1994–2002, Pontificiae Academiae Scientiarum Scripta Varia* 100 (Pontificia Academia Scientiarum, 2003), 336–343, pas.va/content/dam/casinapioiv/pas/pdf-volumi/scripta-varia/sv100pas.pdf.
17. The Latin reads: "*Liber tibi sit paginia divina, ut haec audias; liber tibi sit*

orbis terrarium, ut haec videas." See also "2120 *enarrationes in psalmos,"* Writings of Augustine, augnet.org, accessed June 1, 2023, augnet.org/en/works-of-augustine/writings-of-augustine/2120-enarrationes-in-psalmos/.

18. The martyr William Tyndale (c. 1494–1536) translated the New Testament into English so that multitudes in England could read those Scriptures in their mother tongue.

19. While the media invariably suggest that Edwin Hubble discovered the expanding universe, that is not true.

20. Edith Eva Eger, *The Choice: Embrace the Possible* (Scribner, 2017), 156.

21. See Blaise Pascal, *Thoughts,* trans. W. F. Trotter, in vol. 48 of *The Harvard Classics,* ed. Charles W. Eliot (Collier, 1910), 99. The original French quote reads thus: "Le cœur a ses raisons, que la raison ne connaît point."

Chapter 6: Seeing the Designer in the Lab

1. Richard Dawkins, *The Blind Watchmaker: Why the Evidence of Evolution Reveals a Universe Without Design* (W. W. Norton, 1987), 10.

2. Francis H. Crick and Leslie E. Orgel, "Directed Panspermia," *Icarus* 19, no. 3 (July 1973): 341–346, doi:10.1016/0019-1035(73)90110-3.

3. Colin Barras, "First Americans May Have Been Neanderthals 130,000 Years Ago," *New Scientist,* April 26, 2017, newscientist.com/article/2129042-first-americans-may-have-been-neanderthals-130000-years-ago/; Steven R. Holen et al., "A 130,000-Year-Old Archaeological Site in Southern California, USA," *Nature* 544 (April 27, 2017): 479–483, doi:10.1038/nature22065.

4. Todd J. Braje et al., "Were Hominins in California ~130,000 Years Ago?," *PaleoAmerica* 3, no. 3 (2017): 200–202, doi:10.1080/20555563.2017.1348091.

5. Ruth Gruhn, "Observations Concerning the Cerutti Mastodon Site," *PaleoAmerica* 4, no. 2 (2018): 101–102, doi:10.1080/20555563.2018.1467192.

6. Fazale Rana, *The Cell's Design: How Chemistry Reveals the Creator's Artistry* (Baker Books, 2008), 269–283.

7. William Paley, *Natural Theology or Evidence of the Existence and Attributes of the Deity, Collected from the Appearances of Nature,* Oxford World's Classics, ed. Matthew D. Eddy and David Knight (Oxford University Press, 2006), 7–8.

8. Joel S. Bader et al., "DNA Transport by a Micromachined Brownian Ratchet Device," *Proceedings of the National Academy of Sciences, USA* 96, no. 23 (November 9, 1999): 13165–13169, doi:10.1073/pnas.96.23.13165.

9. Bader et al., "DNA Transport," 13165–13169.

10. Qing Guo and Rui Sousa, "Translocation by T7 RNA Polymerase: A Sensitively Poised Brownian Ratchet," *Journal of Molecular Biology* 358, no. 1 (April 21, 2006): 241–254, doi:10.1016/j.jmb.2006.02.001.

11. Ping Xie, "Model for Forward Polymerization and Switching Transition Between Polymerase and Exonuclease Sites by DNA Polymerase Molecular Motors," *Archives of Biochemistry and Biophysics* 457, no. 1 (January 1, 2007): 73–84, doi:10.1016/j.abb.2006.09.019; Ping Xie, "A Possible Mechanism for the Dynamics of Transition Between Polymerase and Exonuclease Sites in a High-Fidelity DNA Polymerase," *Journal of Theoretical Biology* 259, no. 3 (August 7, 2009): 434–439, doi:10.1016/j.jtbi.2009.04.009; Ping Xie, "A Nucleotide Binding Rectification Brownian Ratchet Model for Translocation of Y-Family DNA Polymerases," *Theoretical Biology and Medical Modelling* 8 (June 24, 2011): id. 22, doi:10.1186/1742-4682-8-22.

Chapter 7: A Vision for Archaeology and Faith

1. Frank Morison, *Who Moved the Stone?* (Faber & Faber; The Century Co., 1930). An independently published 2017 edition with amendments is currently available on Amazon at amazon.com/Who-Moved-Stone-Christian-Classics/dp/1521209677.
2. William Ramsay, *St. Paul the Traveller and the Roman Citizen* (Andesite Press, 2017).
3. Josh McDowell and Sean McDowell, *More Than a Carpenter*, rev. ed. (Tyndale, 2009).
4. This is famously presented in Richard C. Lewontin, "Billions and Billions of Demons," *The New York Review*, January 9, 1997, nybooks.com/articles/1997/01/09/billions-and-billions-of-demons/.
5. Real Seekers Ministries, "Dr. Gary Habermas—Assessing the Pre-NT Christian Creeds," Real Seekers, July 23, 2020, YouTube video, 1:19:45, youtube.com/watch?v=o2IA6i56Elc.
6. Gary Habermas and Michael Licona, *The Case for the Resurrection of Jesus* (Kregel Publications, 2004); Gary Habermas, *Evidence for the Historical Jesus: Is the Jesus of History the Christ of Faith?* (Christian Publishing House, 2020). See also Gary Habermas, *On the Resurrection: Evidences*, vol. 1 (B&H Academic, 2024).
7. Kristin Romey, "Unsealing of Christ's Reputed Tomb Turns Up New Revelations," *National Geographic*, October 30, 2016, nationalgeographic.com/culture/article/jesus-christ-tomb-burial-church-holy-sepulchre.
8. Vassilios Tzaferis, "Crucifixion—The Archaeological Evidence," *Biblical Archaeology Review* 11, no. 1 (January/February 1985): 44–53, baslibrary.org/biblical-archaeology-review/11/1/6. See also, Biblical Archaeology Society Staff, "Roman Crucifixion Methods Reveal the History of Crucifixion," *Bible History Daily* (blog), Biblical Archaeology Society, January 13, 2024, biblicalarchaeology.org/daily/biblical-topics/crucifixion/roman-crucifixion-methods-reveal-the-history-of-crucifixion/; Amanda Borschel-Dan, "How Jesus Died: Extremely Rare Evidence of Roman Crucifixion Uncovered in Italy," *Times of Israel*, May 30, 2018, timesofisrael.com/extremely-rare-archaeological-evidence-of-roman-crucifixion-uncovered-in-italy/; and David Ingham and Corinne Duhig, "Crucifixion in the Fens: Life and Death in Roman Fenstanton," *British Archaeology* (January–February 2022): 18–29, archaeologyuk.org/resource/free-access-to-crucifixion-in-the-fens-life-and-death-in-roman-fenstanton.html.
9. Jack Finegan, *The Archaeology of the New Testament: The Life of Jesus and the Early Church* (Princeton University Press, 1993), 359–374.
10. E. L. Sukenik, "The Earliest Records of Christianity," *American Journal of Archaeology* 51, no. 4 (October–December 1947): 351–365, jstor.org/stable/500006.
11. "The Dead Sea Scrolls," Israel Museum, Jerusalem, accessed April 8, 2024, imj.org.il/en/wings/shrine-book/dead-sea-scrolls.
12. "The Great Isaiah Scroll," The Digital Dead Sea Scrolls, Israel Museum, Jerusalem, accessed April 8, 2024, dss.collections.imj.org.il/isaiah.
13. Eilat Mazar, "Is This the Prophet Isaiah's Signature?," *Biblical Archaeology Review* 44, no. 2 (March/April and May/June 2018), 64–73, 92, baslibrary.org/biblical-archaeology-review/44/2/7.
14. Nigel Reynolds, "Tiny Tablet Provides Proof for Old Testament," *Telegraph*, July 11, 2007, telegraph.co.uk/news/uknews/1557124/Tiny-tablet-provides-proof-for-Old-Testament.html. See also Bryant Wood, "Nebo-Sarsekim Found in Babylonian Tablet," Associates for Biblical Research, April 28, 2008, biblearchaeology.org/research/contemporary-issues/3520-nebosarsekim-found-in-babylonian-tablet.
15. J. Warner Wallace, *Cold-Case Christianity: A Homicide Detective Investigates the Claims of the*

Gospels (David C. Cook, 2013).

16. For further reading, I recommend the following books by Titus Kennedy: *Unearthing the Bible: 101 Archaeological Discoveries That Bring the Bible to Life* (Harvest House, 2020); *Excavating the Evidence for Jesus: The Archaeology and History of Christ and the Gospels* (Harvest House, 2022); and *The Essential Archaeological Guide to Bible Lands: Uncovering Biblical Sites of the Ancient Near East and Mediterranean World* (Harvest House, 2023).

Chapter 8: Why Science Was Stifled in China

1. Charles Cotherman, "Reflections on the History and Legacy of *Radix*," *Radix Magazine*, September 12, 2020, radixmagazine.com/2020/09/12/reflections-on-the-history-and-legacy-of-radix/.
2. See, for example, Walter R. Hearn, *Being a Christian in Science* (InterVarsity Press, 1997).
3. J. C. Mather et al., "A Preliminary Measurement of the Cosmic Microwave Background Spectrum by the Cosmic Background Explorer (COBE) Satellite," *Astrophysical Journal Letters* 354 (May 1990): L37, doi:10.1086/185717.
4. Joseph Needham, *Science and Civilisation in China Series*, 7 vols. (Cambridge University Press, 1957).
5. For a brief introduction, see chapter 7 in Kenneth R. Samples, *God Among Sages* (Baker Books, 2017).
6. *ren- ai* (仁爱) is the Confucian concept of benevolence or compassion. It can be translated *agape*, though it does not carry the Christian meaning of sacrificial love.
7. Confucius, 《論語. 雍也》 "敬鬼神而远之," in *The Analects of Confucius: An Online Teaching Translation*, trans. R. Eno, version 2.21 (2015), https://shorturl.at/bnaVr, Book VI, Item 6.22, p. 27.
8. Sik-Pui Wong, *Sacrificial Love—Portraits of CIM Missionaries* (Chinese language), (CCM Publishers, 2006); 黃錫培 《舍命的爱 – 中国内地会宣教士小传》 (美国中信出版社, 2006).
9. Wikipedia, s.v. "Zheng He," last modified May 26, 2023, en.wikipedia.org/wiki/Zheng_He.
10. The nine social echelons are further divided into the upper, middle, and lower categories. In the upper category are the following professions: kings, sages, hermits, immortals, literary scholars, samurai, peasants, laborers/artisans, and merchants. In the middle category are pre-scholars, doctors, fortune-tellers, artists, students, entertainers, monks, Taoist priests, and nuns. In the lower category are witches, showgirls/prostitutes, medicine-men, watch-men, barbers, musicians, actors, paupers, and candy men/hawkers.
11. Most Chinese consider Buddhism their indigenous religion, though it originated in India and spread to China and other Asian countries around the first century AD.
12. 天人合一 [English translation: "Heaven and man are one."]
13. The Tsinghua University motto is derived from the Taoist "Yi Jing" (*Book of Change*): 《周易》"天行建，君子以自强不息"（乾卦）、"地势坤，君子以厚德载物"（坤卦）. [English translation: "As the heavens maintain vigor through movement, the gentleman should constantly strive for self-perfection. As the Earth is vast, the gentleman should have the breadth of character to carry different circumstances."]
14. This is a lesson in missiology as some missionaries leveraged this "opportunity" to enter China, and earned the black mark that Christianity is an imperialistic arm of the West.
15. See, for example: In-Sing Leung, *Reflection on Mystery of the Universe* [Chinese language], rev. ed. (CCM Publishers, 1996); 梁燕城 《静思宇宙玄秘》 （中信出版社, 1996年增订版.

16. "Peking Man," *World Archaeology*, no. 48 (July 3, 2011), world-archaeology.com/great-discoveries/great-discoveries-peking-man/.
17. Fazale Rana, "Were They Real? The Scientific Case for Adam and Eve," Reasons to Believe, October 1, 2010, reasons.org/explore/publications/articles/were-they-real-the-scientific-case-for-adam-and-eve.
18. Li Jin and Bing Su, "Natives or Immigrants: Origin and Migrations of Modern Humans in East Asia," *Nature Reviews Genetics* 1 (2000): 126–133, doi:10.1038/35038565; "Chinese Ancestors Were Africans Leading Geneticist Says," posted by Khalif Sarahize, May 14, 2013, YouTube video, 2:09, youtube.com/watch?v=TzyoMqxppR0.

Chapter 9: Reflections from a "Rocket Girl"

1. Francis Thompson, "The Hound of Heaven," 1890, houndofheaven.com/poem. The "Hound of Heaven" is metaphor for the Holy Spirit, coined by Thompson and later used by C. S. Lewis as he describes his reluctant conversion to Christianity.
2. For further reading about this evidence, see Leslie Wickman, *God of the Big Bang: How Modern Science Affirms the Creator* (Worthy Books, 2015).
3. Leslie Wickman, "CNN Op-Ed on Discovery of Gravity Waves," Starry Nights, Inc., *Reflections* (blog), March 31, 2014, starrynights.me/reflections/2014/3/31/cnn-op-ed-on-discovery-of-gravity-waves.html. Originally published as Leslie Wickman, "Does Big Bang Breakthrough Offer Proof of God?," CNN, *Belief Blog*, March 20, 2014.
4. Leslie Wickman, "Is the Conflict Between Science and Religion Real?," TEDx, Azusa Pacific University, YouTube video, 18:00, posted by TEDx Talks, May 19, 2015, accessed April 9, 2024, youtube.com/watch?v=xGbW-tYn6ZA.
5. See Mark Ross, "In Essentials, Unity, in Non-Essentials, Liberty, in All Things, Charity," *Ligonier*, September 1, 2009, ligonier.org/learn/articles/essentials-unity-non-essentials-liberty-all-things.

Chapter 10: Is There Compatibility Between the Big Bang and the Bible?

1. There are many objective reasons to believe that the Bible provides an accurate historical record, including the archaeological discoveries that have substantiated its historical accuracy. However, it is not the purpose of this chapter to describe the objective evidence for such a conclusion. Some details can be found in chapter 7, "A Vision of Archaeology and Faith," by John Bloom.
2. The first Hugh Ross book that I read was *The Fingerprint of God*, followed by *The Creator and the Cosmos*, and then *A Matter of Days*.
3. Edwin Hubble, "A Relation Between Distance and Radial Velocity Among Extra-Galactic Nebulae," *Proceedings of the National Academy of Sciences USA* 15, no. 3 (April 15, 1929): 168–173, doi:10.1073/pnas.15.3.168.
4. A. A. Penzias and R. W. Wilson, "A Measurement of Excess Antenna Temperature at 4800 Mc/s," *Astrophysical Journal Letters* 142 (July 1965): 419–421, doi:10.1086/148307; R. H. Dicke et al., "Cosmic Black-Body Radiation," *Astrophysical Journal Letters* 142 (July 1965): 414–419, doi:10.1086/148306.
5. Stephen Hawking and G. F. R. Ellis, *The Large-Scale Structure of Space-Time* (Cambridge University Press, 1973); Roger Penrose, "Gravitational Collapse and Space-Time Singularities," *Physical Review Letters* 14, no. 57 (January 1965), doi:10.1103/PhysRevLett.14.57.
6. Arvind Borde, Alan H. Guth, and Alexander Vilenkin, "Inflationary Spacetimes Are Incomplete in Past Directions," *Physical Review Letters* 90, nos. 15–18 (April 18, 2003): 151301,

doi:10.1103/PhysRevLett.90.151301.

7. See, for example, Peter D. Ward and Donald Brownlee, *Rare Earth: Why Complex Life is Uncommon in the Universe* (Copernicus, 2000), 202.
8. W. D. Mounce, ed., *Mounce's Complete Expository Dictionary of Old and New Testament Words* (Zondervan, 2006), 157–158.
9. Gleason Archer, *A Survey of Old Testament Introduction*, 3rd rev. and exp. ed. (Moody Press, 1994), 199.
10. *Strong's Exhaustive Concordance of the Bible* (Abingdon Press, 1890) lists 8,674 Hebrew and Chaldean words. Rodney Whitefield, *Reading Genesis One* (R. Whitefield, 2003), cites Robert L. Harris, Gleason L. Archer Jr., and Bruce K. Waltke, *Theological Wordbook of the Old Testament* (Moody, 1980), as stating there are 2,552 root words in Hebrew Scripture.
11. The Hebrew word `owlam is the only word that could possibly mean "era," but it almost always refers to an eternal time period, a time period from eternity, or a perpetual time period. The King James Version Old Testament lexicon says `owlam is used 439 times. The KJV translates it as "long" only two times. Its English translations and the number of times it is translated are listed as the following: "ever" 272 times; "everlasting" 63 times; "old" 22 times; "perpetual" 22 times; "evermore" 15 times; "never" 13 times; "time" 6 times; "ancient" 5 times; "world" 4 times; "always" 3 times; "alway" (archaic for "always") 2 times; "long" 2 times; "more" 2 times; "never" + `al 2 times; and miscellaneous translations 6 times.
12. John Sailhamer, *Genesis*, in *The Expositor's Bible Commentary*, ed. Tremper Longman III and David E. Garland, vol. 1, *Genesis–Leviticus*, rev. ed. (Zondervan, 2008), 59.
13. Ward and Brownlee, *Rare Earth*, 202.
14. Sailhamer, *Genesis*, 64.
15. The King James Version Hebrew lexicon defines *nephesh* as "soul, self, life, creature, person, appetite, mind, living being, desire, emotion, passion." Note the close association between this word and the "soulish" aspect of higher mammals, which includes appetite, desire, mind, emotion, and passion.
16. Malcolm Browne, "Clues to the Universe's Origin Expected," *New York Times*, March 12, 1978, p. 1, col. 54.

Chapter 11: Mission Accomplished

1. C. S. Lewis, *Mere Christianity* (HarperCollins, 1952), 63.

Chapter 12: God's "Two Books" Still Speak

1. Fred Hoyle, *The Nature of the Universe* (Basil Blackwell, 1952), 109.
2. Hoyle, *Nature of the Universe*, 109.
3. Immanuel Kant, *The Critique of Pure Reason* in *Great Books of the Western World*, vol. 42, ed. Robert Maynard Hutchins (Encyclopedia Britannica, 1952).
4. Kant, *The Critique of Pure Reason*, 134–138.
5. Stephen W. Hawking and George F. R. Ellis, "The Cosmic Black-Body Radiation and the Existence of Singularities in Our Universe," *Astrophysical Journal* 152 (April 1968): 25–36, doi:10.1086/149520; Stephen William Hawking and Roger Penrose, "The Singularities of Gravitational Collapse and Cosmology," *Proceedings of the Royal Society of London A* 314 (January 1970): 529–548, doi:10.1098/rspa.1970.0021.
6. For a detailed description and explanation, see my book *Navigating Genesis* (RTB Press, 2013), chaps. 4–6.
7. L. Paul Knauth and Marin J. Kennedy, "The Late Precambrian Greening of the Earth," *Nature*

460, no. 7256 (August 6, 2009): 728–732, doi:10.1038/nature08213; Paul K. Strother et al., "Earth's Earliest Non-Marine Eukaryotes," *Nature* 473, no. 7348 (May 26, 2011): 505–509, doi:10.1038/nature09943.

8. Hugh Ross, *Designed to the Core* (RTB Press, 2022), 178–181.

9. Hugh Ross, "Benefits of Viruses," *Today's New Reason to Believe* (blog), Reasons to Believe, August 1, 2022, reasons.org/explore/blogs/todays-new-reason-to-believe/benefits-of-viruses.

10. Ross, *Designed to the Core*, 97–101.

Index

Contributors

David L. Block
David Block is professor emeritus in the School of Computer Science and Applied Mathematics at the University of the Witwatersrand, Johannesburg, South Africa. David was elected a Fellow of the Royal Astronomical Society of London at the age of 19 and has been a visiting research astronomer at the Australian National University (ANU), the European Southern Observatory in Germany, the California Institute of Technology, and Harvard University.

John A. Bloom
John Bloom earned a PhD in physics from Cornell University, then went to seminary and completed a second PhD in Ancient Near Eastern studies (Old Testament backgrounds) so that he would gain advanced formal training in both science and theology. For the past 29 years he has taught physics at Biola University. He founded the Master of Arts in Science and Religion program at Biola's Talbot School of Theology in 2004, in collaboration with the Christian Apologetics program.

Krista Bontrager
Krista Bontrager is a fourth-generation minister. She is a Bible scholar, lay minister, author, teacher, former university professor, and homeschool mom. She earned an MA in theology and an MA in Bible exposition from Talbot School of Theology. She is working on a doctor of ministry in apologetics at Birmingham Theological Seminary and is the cofounder of the Center for Biblical Unity. She is also a popular YouTube teacher and podcaster.

Cynthia Cheung
Cynthia Cheung earned her PhD from the University of Maryland and a BA

from University of California, Berkeley, both in astronomy. She worked for 35 years in various roles for NASA space missions, gaining significant experience in space science mission operations. An astrophysicist by training, her research interests include high energy astrophysics, stellar evolution, and nucleosynthesis. Cynthia has served in multiple leadership positions in college fellowships and Chinese churches in the US, giving science and faith seminars in both English and Chinese.

Christina A. Cirucci

Christina Cirucci earned a BS in mechanical engineering from Virginia Tech. She worked in engineering for seven years before pursuing a career in medicine. She earned her MD from Jefferson Medical College, then completed a residency in obstetrics and gynecology at the Medical College of Virginia. Christina is board certified in obstetrics and gynecology and has worked for 20 years in the field. Additionally, she has served on many volunteer mission trips to countries in Asia and Africa, including six to Bangladesh.

Francisco Delgado

Francisco Delgado is a medical doctor with a specialty in infectious diseases. He earned his MD from Universidad La Salle in Mexico City and completed his infectious diseases fellowship at Vanderbilt University. He has been in clinical practice for over 25 years. He and his wife live in Indianapolis, Indiana.

George R. Haraksin II

Philosopher-ethicist George Haraksin serves as Scholar Community program director and a staff scholar at Reasons to Believe (RTB). He holds a BA in comparative religions from California State University, Fullerton; an MA in philosophy of religion and ethics from Talbot School of Theology, Biola University; and an MA in philosophy from Claremont Graduate University. George serves as a senior adjunct professor in philosophy and ethics at Azusa Pacific University, where he has taught since 2002.

Balajied Nongrum

Balajied Nongrum serves as the lead consultant of research and apologetics for Reasons to Believe Asia-Pacific (RTB APAC). He holds a bachelor's degree in veterinary science from the Acharya N. G. Ranga Agricultural University, Hyderabad. After completing his master's degree in philosophy and religion, he went on to pursue a second MA in science and religion from Talbot School of Theology. For the past 15 years Balajied has served as both a speaker and trainer on diverse subjects, including ethics, philosophy, science, and religion.

Fazale "Fuz" Rana

Fuz Rana is president, CEO, and senior scholar at Reasons to Believe (RTB). He earned a PhD in chemistry with an emphasis in biochemistry from Ohio University. Fuz conducted postdoctoral work at the Universities of Virginia and Georgia and worked for seven years as a senior scientist in product development for Procter & Gamble. He has published articles in peer-reviewed scientific journals, delivered presentations at international scientific meetings, and addressed the relationship between science and Christianity at churches and universities in the US and abroad. Since joining RTB in 1999, Fuz has participated in numerous podcasts and videos, authored countless articles, and published several books, including *Humans 2.0* (coauthored with Kenneth Samples), *Creating Life in the Lab*, and *Fit for a Purpose*.

David Rogstad

Dave Rogstad earned a BS in physics from the California Institute of Technology (Caltech), followed by an MS and a PhD, also in physics, from Caltech. After leaving Caltech, Dave pursued a 30-year career with the Jet Propulsion Laboratory (JPL), where he worked on several projects (including the Galileo mission and hypercube concurrent computation) and published several professional articles. Dave received numerous awards for his technical contributions to the NASA program, including a NASA Exceptional Service Medal for his contributions to antenna arraying. In 2000, Dave retired from full-time work at JPL to devote himself to ministry at RTB, where he served faithfully as executive vice president and scholar emeritus. In June 2023, God called Dave home to his eternal reward.

Hugh Ross

Hugh Ross is a senior scholar and founder of Reasons to Believe. He earned a BA in physics from the University of British Columbia and a PhD in astronomy from the University of Toronto. As a postdoctoral fellow at the California Institute of Technology he continued his research on quasars and galaxies. After five years there, he transitioned to full-time ministry. His writings include journal and magazine articles, blogs, and numerous books—*The Creator and the Cosmos*, *Why the Universe Is the Way It Is*, and *Designed to the Core*, among others. He has spoken on hundreds of university campuses as well as at conferences and churches around the world.

Michael G. Strauss

Mike Strauss earned his PhD at UCLA, where he studied elementary particle physics. He is an assistant professor of physics at the University of Oklahoma. Mike has conducted research at the Fermi National Accelerator Laboratory (Fermilab) and at CERN. He is currently involved in research measuring properties of the Higgs boson and looking for evidence of nonstandard model Higgs bosons. He is the author of *The Creator Revealed: A Physicist Examines the Big Bang and the Bible*, and one of the general editors of Zondervan's *Dictionary of Christianity and Science*.

Leslie Wickman

Leslie Wickman is a research scientist, engineering consultant, author, and inspirational speaker. Leslie was an engineer for Lockheed Missiles & Space Company for more than a decade, during which she worked on NASA's Hubble Space Telescope and International Space Station programs, receiving commendations from NASA for her contributions and being designated as Lockheed's Corporate Astronaut. She now focuses her time on her nonprofit, Starry Nights, Inc.

Jeff Zweerink

Astrophysicist Jeff Zweerink is a senior research scholar at Reasons to Believe. He earned a PhD in astrophysics with a focus on gamma rays from Iowa State University. Jeff was involved in research projects such as STACEE and VERITAS and is coauthor of more than 30 academic papers. His books include *Escaping the Beginning?*, *Is There Life Out There?*, *Who's Afraid of the Multiverse?*, and *Building Bridges* (coauthor). Jeff writes and speaks on the compatibility of science and the Christian faith and works on GAPS (a balloon experiment seeking to detect dark matter).

About Reasons to Believe

Reasons to Believe (RTB) exists to reveal God in science. Based in Covina, California, RTB was established in 1986 and since then has taken scientific evidence for the God of the Bible across the US and around the world. Our ongoing work is providing content for all who desire to explore the connection between science and the Christian faith.

RTB is unique in its range of resources. The curious can explore articles, podcasts, and videos. Those who want to learn more can delve into books, in-person and livestreamed events, and online courses. Donors enable us to continue this important work.

For more information, visit reasons.org.

For inquiries, contact us via:
818 S. Oak Park Rd.
Covina, CA 91724
(855) REASONS | (855) 732-7667
ministrycare@reasons.org

 @rtb_official

Test ideas and sharpen your thinking skills.

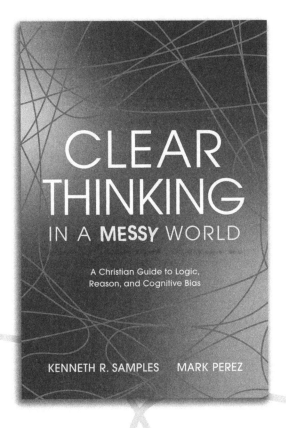

Scan here

reasons.org/think

"Kenneth Samples and Mark Perez have created a superb book....
It's highly accessible and strikes the perfect balance of logic and
critical thinking, all with an apologetic flavor that trains you to
think better."

—Brian Auten, Founder of Apologetics 315
and host of the *Apologetics 315* podcast

Discover easy-to-understand responses to common questions about Scripture and science, including:

- Is there conflict between science and biblical creation?
- Are the big bang and evolution inseparable?
- Can the fossil record be viewed through a scriptural lens?

Download the free ebook today and gain confidence that the Christian faith not only withstands curiosity—it welcomes it.

reasons.org/harmony